Springer Theses

Recognizing Outstanding Ph.D. Research

Aims and Scope

The series “Springer Theses” brings together a selection of the very best Ph.D. theses from around the world and across the physical sciences. Nominated and endorsed by two recognized specialists, each published volume has been selected for its scientific excellence and the high impact of its contents for the pertinent field of research. For greater accessibility to non-specialists, the published versions include an extended introduction, as well as a foreword by the student’s supervisor explaining the special relevance of the work for the field. As a whole, the series will provide a valuable resource both for newcomers to the research fields described, and for other scientists seeking detailed background information on special questions. Finally, it provides an accredited documentation of the valuable contributions made by today’s younger generation of scientists.

Theses are accepted into the series by invited nomination only and must fulfill all of the following criteria

- They must be written in good English.
- The topic should fall within the confines of Chemistry, Physics, Earth Sciences, Engineering and related interdisciplinary fields such as Materials, Nanoscience, Chemical Engineering, Complex Systems and Biophysics.
- The work reported in the thesis must represent a significant scientific advance.
- If the thesis includes previously published material, permission to reproduce this must be gained from the respective copyright holder.
- They must have been examined and passed during the 12 months prior to nomination.
- Each thesis should include a foreword by the supervisor outlining the significance of its content.
- The theses should have a clearly defined structure including an introduction accessible to scientists not expert in that particular field.

More information about this series at http://www.springer.com/series/8790

Son Duy Dao

Modelling and Intelligent Optimisation of Production Scheduling in VCIM Systems

Doctoral Thesis accepted by
University of South Australia, Adelaide, Australia

Springer

Author
Dr. Son Duy Dao
School of Engineering
University of South Australia
Adelaide, SA
Australia

Supervisors
Prof. Kazem Abhary
School of Engineering
University of South Australia
Adelaide, SA
Australia

Dr. Romeo Marian
School of Engineering
University of South Australia
Adelaide, SA
Australia

ISSN 2190-5053 ISSN 2190-5061 (electronic)
Springer Theses
ISBN 978-3-319-72112-5 ISBN 978-3-319-72113-2 (eBook)
https://doi.org/10.1007/978-3-319-72113-2

Library of Congress Control Number: 2017960925

Printed on acid-free paper

This Springer imprint is published by Springer Nature
The registered company is Springer International Publishing AG
The registered company address is: Gewerbestrasse 11, 6330 Cham, Switzerland

I dedicate this Ph.D. to my parents! My parents used to be the poor farmers and did not have high school degrees, but they were still "smart" enough to understand the value of education. My parents always made my education the first priority because they strongly believed my education would change the poor living condition of my family. My parents had to work very hard and borrow lots of money to send me to a university in Vietnam. Their love and belief made me stronger to study. As a village boy, I struggled with the university work in the first year; but I did much better later on. As a result, in 2004, I was the first one in my family, who got a university degree.
I graduated with a good grade and with two medals (silver and bronze) in the National Student Olympic Competition in Mechanics, annually organised by Vietnam Ministry of Education and Training. These initial achievements changed my life and my family forever.

I also dedicate this Ph.D. to my wife Nguyễn Thị Mai Anh and my daughter Đào Thị Minh Khuê, who have endured the absence of a husband and a father, with patience and love, during my Master and Ph.D. studies in Australia.

Supervisor's Foreword

Virtual computer-integrated manufacturing (VCIM) is a new manufacturing system, which is being developed to help manufacturing enterprises survive and thrive in today's global market. VCIM is a global manufacturing system, capable of effectively exploiting manufacturing resources that are locally and/or globally distributed. Production scheduling plays an important role in VCIM systems. Currently, there are two major limitations in the VCIM production scheduling, namely lack of a comprehensive scheduling model and lack of a robust optimisation method.

To overcome these two limitations, an innovative VCIM production scheduling model and a robust genetic algorithm (GA) have been developed in this thesis. The developed model is the most comprehensive VCIM production scheduling model to date, which is capable of not only simultaneously supporting collaborative shipment scheduling and handling multiple product orders but also dealing with multiple objective functions under uncertainties. The developed GA with innovative algorithm structure, chromosome encoding, crossover and mutation is capable of searching for optimal/suboptimal solutions to large-scale complex scheduling problems in the developed VCIM production scheduling model.

The effectiveness of the developed VCIM production scheduling model as well as the developed GA has been demonstrated via a comprehensive case study, i.e. a VCIM system producing electric bikes. In this case study, the superiority and advantage of the developed model and GA are verified by comparing the former with two other models and the latter with three well-known optimisation algorithms, namely traditional GA, cuckoo search, and particle swarm optimisation. The outcome of this research can serve as a foundation towards developing a decision support system capable of helping decision-makers to operate VCIM systems more effectively.

It has been a great pleasure for me to supervise the author of this thesis, Dr. Son Duy Dao, who was one of the top Ph.D. students I have ever had. The outstanding quality of his thesis was confirmed by two international examiners who passed the thesis with just minor corrections. From this thesis, 14 publications were produced: ten peer-reviewed journal articles, one peer-reviewed book chapter, three peer-reviewed conference papers; two more journal articles are currently under review for publications. As a consequence of the exceptional quality of this research, the author won a prestigious international student award "*Academic Excellence in Postgraduate Research*" conferred by the Governor of South Australia, Australia, in 2016.

Adelaide, Australia Prof. Kazem Abhary

Parts of this thesis have been published in the following documents:

Peer-reviewed Journal Articles

1. **Dao S.D.**, Abhary K., Marian R., 2014, "Optimisation of partner selection and collaborative transportation scheduling in virtual enterprises using GA", *Expert Systems with Applications*, vol. 41, no. 15, pp. 6701–6717. (**Elsevier**)
2. **Dao S.D.**, Abhary K., Marian R., 2016, "An innovative model for resource scheduling in VCIM systems", *Operational Research: An International Journal*, dOI: 10.1007/s12351-016-0252-y. (**Springer**)
3. **Dao S.D.**, Abhary K., Marian R., 2016, "An improved structure of genetic algorithms for global optimisation", *Progress in Artificial Intelligence*, vol. 5, no. 3, pp. 155–163. (**Springer**)
4. **Dao S.D.**, Abhary K., Marian R., 2016, "Maximising performance of genetic algorithm solver in Matlab", *Engineering Letters*, vol. 24, no. 1, pp. 75–83. (**IAENG**)
5. **Dao S.D.**, Abhary K., Marian R., 2016, "A stochastic production scheduling model for VCIM systems", *Intelligent Industrial Systems*, vol. 2, no. 1, pp. 85–101. (**Springer**)
6. **Dao S.D.**, Abhary K., Marian R., 2017, "An integrated production scheduling model for multi-product orders in VCIM systems", *International Journal of System Assurance Engineering and Management*, vol. 8, no. 1, pp. 12–27. (**Springer**)
7. **Dao S.D.**, Abhary K., Marian R., 2017, "An improved genetic algorithm for multidimensional optimisation of precedence-constrained production planning and scheduling", *Journal of Industrial Engineering International,* dOI: 10.1007/s40092-016-0181-7. **(Springer)**
8. **Dao S.D.**, Abhary K., Marian R., 2017, "Optimisation of assembly scheduling in VCIM systems using genetic algorithm", *Journal of Industrial Engineering International,* dOI: 10.1007/s40092-017-0183-0. **(Springer)**
9. **Dao S.D.**, Abhary K., Marian R., 2017, "A bibliometric analysis of genetic algorithms throughout the history", *Computers & Industrial Engineering,* vol. 110, pp. 395–403. (**Elsevier**)
10. **Dao S.D.**, Abhary K., Marian R., 2017, "An innovative framework for designing genetic algorithm structures", *Expert Systems with Applications*, vol. 90, pp. 196–208. (**Elsevier**)

Peer-reviewed Book Chapter

11. **Dao S.D.**, Abhary K., Marian R., 2017, "The latest developments in virtual computer-integrated manufacturing systems", *IAENG Transactions on Engineering Sciences,* World Scientific, DOI: 10.1142/9789813226203_0019, pp. 232–239.

Peer-reviewed Conference Papers

12. **Dao S.D.**, Abhary K., Marian R., 2015, "An adaptive restarting genetic algorithm for global optimisation", *Proceedings of the World Congress on Engineering and Computer Science*, pp. 455-459, 21–23 October, San Francisco, USA.
13. **Dao S.D.**, Abhary K., Marian R., 2016, "Is a virtual computer-integrated manufacturing system feasible?", *Proceedings of the World Congress on Engineering*, pp. 733–736, 29 June–1 July, London, U.K.
14. **Dao S.D.**, Abhary K., Marian R., 2016, "An effective genetic algorithm for large-scale traveling salesman problems", *Proceedings of the World Congress on Engineering and Computer Science*, pp. 516-520, 19–21 October, San Francisco, USA.

Peer-reviewed Journal Articles (Under Review)

15. **Dao S.D.**, Abhary K., Marian R., 2017, "A modified genetic algorithm for optimisation of collaborative manufacturing-transportation scheduling in VCIM systems", *Operational Research: An International Journal.* (**Springer**)
16. **Dao S.D.**, Abhary K., Marian R., 2017, "Optimisation of production scheduling for multi-product orders in VCIM systems using GA", *Computers & Industrial Engineering.* **(Elsevier)**

Acknowledgements

I would like to express the heartfelt gratitude to my supervisors, Prof. Kazem Abhary and Dr. Romeo Marian, at School of Engineering, University of South Australia (UniSA) in Australia, from whom not only I received very professional supervision and guidance for my research but also I learned invaluable lessons for academic life. I really appreciate the encouragement, trust and freedom that my Ph.D. supervisors gave me to conduct the research of my own interest.

I would like to thank Thai Nguyen University of Technology (TNUT) in Vietnam, where I studied and have been working for. I would like to thank Prof. Tuan Van Nguyen at TNUT in Vietnam for his support during the time I did bachelor's degree. I would like to thank Vietnamese Government for sponsoring my Master study at UniSA in Australia.

I am so grateful to Australian Government for sponsoring my Ph.D. research at UniSA in Australia in the form of Endeavour scholarship. I would like to thank my Endeavour case managers for their excellent service. My sincere thanks are extended to the staff members, librarians and student fellows at UniSA in Australia for their kind support and assistance throughout my candidature. In addition, I would like to thank the StudyAdelaide and the Governor of South Australia for the 2016 international student award, "*Academic Excellence in Postgraduate Research*".

Contents

Abbreviations

AGV	Automated Guided Vehicle
AITS	Automated Inspection and Testing System
AMHS	Automated Material Handling System
AMT	Advanced Manufacturing Technology
ANOVA	Analysis of Variance
B2B	Business-to-Business
B2C	Business-to-Customer
C2C	Customer-to-Customer
CAD	Computer-Aided Design
CAE	Computer-Aided Engineering
CAM	Computer-Aided Manufacturing
CAPP	Computer-Aided Process Planning
CIM	Computer-Integrated Manufacturing
CM	Cellular Manufacturing
CNC	Computer Numerical Control
DPMO	Defects Per Million Opportunity
FMC	Flexible Manufacturing Cell
FMS	Flexible Manufacturing System
GA	Genetic Algorithm
JIT	Just-In-Time
MRP	Material Requirement Planning
NP	Nondeterministic Polynomial time
PLC	Programmable Logic Controller
SMEs	Small- and Medium-sized Enterprises
VCIM	Virtual Computer-Integrated Manufacturing

Chapter 1
Introduction

1.1 Introduction

This research is about modelling and genetic algorithm optimisation of production scheduling in VCIM systems. The research has two main parts, namely developing a new production scheduling model for VCIM systems and developing a robust GA to solve the optimisation problem in the developed model. In this chapter, background for the main theme of the research is presented. Motivations are then given to highlight the significance of the research. Next, research objectives and research scope are set. Finally, the thesis structure is presented.

1.2 Research Background

In order to succeed in highly competitive global market today, manufacturing enterprises need to be able to provide higher quality services with lower cost in shorter time. To do so, the enterprises must be innovative, flexible, adaptable, and responsive to changes. These requirements have forced a large number of manufacturing enterprises to apply advanced manufacturing technologies in various types to improve their performances (Gunawardana 2006). In general, Advanced Manufacturing Technology (AMT) is defined as technology associated with software and hardware of computer and numerical based apparatus which are designed to accomplish or support manufacturing tasks (Costa et al. 2000).

Over the past decades, a large number of AMTs have been developed such as Computer Aided Design (CAD), Computer Numerical Control (CNC), Programmable Logic Controllers (PLC), Computer Aided Manufacturing (CAM), Computer Aided Engineering (CAE), Computer Aided Process Planning (CAPP), Automated Guided Vehicles (AGV), Cellular Manufacturing (CM), Material Requirement Planning (MRP), Manufacturing Resource Planning (MRP II),

S. D. Dao, *Modelling and Intelligent Optimisation of Production Scheduling in VCIM Systems*, Springer Theses, https://doi.org/10.1007/978-3-319-72113-2_1

Automated Inspection and Testing Systems (AITS), Automated Material Handling Systems (AMHS), Just-in-time (JIT), Flexible Manufacturing Cells (FMC), Flexible Manufacturing Systems (FMS), Computer Integrated Manufacturing (CIM), etc. (Costa et al. 2000; Gunawardana 2006; Koc and Bozdag 2009; Small and Yasin 1997).

There are a number of criteria to classify the AMTs such as the level of integration, functional application, nature of apparatus, level of organisational integration, and imbedded information processing capabilities (Costa et al. 2000). Based on the degree of integration, AMTs are classified into three levels: *stand-alone level* such as CAD, CNC or CAPP, *intermediate level* such as MRP II, AITS or AMHS, and *integrated level* such as FMC, FMS or CIM (Small and Yasin 1997; Suresh and Meredith 1985). As a fully integrated system, CIM has a significant potential in modern manufacturing industry (Nagalingam and Lin 1999).

A CIM system is a manufacturing system in which computers are used to control the production processes. All functional areas in the system such as design, planning, analysis, purchasing, inventory control, cost accounting, production, marketing, material handling, distribution or management are linked together by means of computer systems. Therefore, the manufacturing can be automated, faster and less error-prone (Miller et al. 2010). Effective integration of AMTs in various functional units makes CIM systems very powerful (Nagalingam and Lin 1999). However, a CIM system can only exploit the local manufacturing resources and its capability will be limited if the resources are not always available (Wang 2007).

Because of globalisation and advances in computer as well as information technology, CIM systems have been evolving to the next stage: Virtual Computer-Integrated Manufacturing (VCIM) systems. VCIM is a relatively new concept, defined as a network of interconnected CIM systems which are locally and/or globally distributed (Nagalingam and Lin 1999). The concept of VCIM has been proposed for one and a half decade with purpose of overcoming the limitation of traditional CIM systems since they only work within one enterprise. Generally speaking, VCIM concept is a combination of CIM and Virtual Enterprise concepts. Virtual Enterprise is a temporary alliance of enterprises, working together to share skills, competencies and resources to better respond to business opportunities (Camarinha-Matos and Afsarmanesh 1999). VCIM inherits the good characteristics and addresses the limitations of CIM and Virtual Enterprise. That is why VCIM systems have two special characteristics, "integration" and "temporary cooperation".

VCIM is still being conceptually developed (Dao et al. 2016a). The latest definition of VCIM is as follows. VCIM is a network of manufacturing enterprises, in which the enterprise members temporarily work together in an integrated manner to create a temporary production system to fulfil the customer order. This temporary production system and temporary co-operation will dissolve after the order is fulfilled (Zhou et al. 2010). It should be noted that a new temporary production system will be created for the next customer order by a group of the same and/or different enterprise members. In addition, manufacturing enterprises in a VCIM system are locally and/or globally distributed. Of course, a lot of work needs to be done to make VCIM systems a reality, e.g. information sharing, security, copyright, membership, quality control, etc. A real VCIM system does not exist yet, although some theoretical work has been done. Nevertheless, inspired by the big success of Alibaba model (Liu and Avery 2009; Waite 2014), VCIM systems are likely to be promising solutions for many small and medium-size manufacturing enterprises in near future, because VCIM systems are capable of exploiting distributed resources effectively, both locally and globally.

Production scheduling is very important to VCIM systems because it will affect the quality, cost as well as lead time of products. After receiving product orders, the VCIM system decomposes the orders into a number of pre-defined standard product components or subassemblies (called *components*, for short, hereafter) which can be independently produced; and then it does the production scheduling by (1) selecting some component suppliers to produce the required components, (2) selecting some assembly agents to assemble the required products and (3) scheduling the shipments to transport the components and products to the required destinations; so that a temporary VCIM production system can be formed to fulfil the customer orders. The whole procedure will be repeated for the next product orders. VCIM production scheduling is a very complex problem, which requires a comprehensive VCIM production scheduling model and a robust optimisation solution method. This research will address these two issues.

1.3 Research Motivations

VCIM systems could help many small and medium-size enterprises (SMEs) survive and thrive in today's competitive global market. SMEs, a symbol of innovation and competition in various industrial sectors, play an important role in any economy (Angilella and Mazzù 2015; Kumari et al. 2015). In today's global market, SMEs face fierce competition from large enterprises that have huge resources. To deal with the competition, SMEs usually concentrate on their core competence and outsource some non-core activities (Kumari et al. 2015). Cooperation among SMEs worldwide to exploit the distributed resources is a critical strategy to enhance their competitiveness (Iturrioz et al. 2015). Such cooperation is perfectly supported by VCIM systems.

Inspired by the big success of Alibaba model (Liu and Avery 2009; Waite 2014), VCIM systems are likely to be promising solutions for many SMEs in near future. Although a real VCIM system does not exist yet, building a VCIM system is feasible because Alibaba.com platform can support its implementation (Dao et al. 2016b). Alibaba.com is a great platform that enables SMEs to do business efficiently anywhere in the world (Yazdanifard and Li 2014). The philosophy that makes Alibaba Company so successful is as follows:

> Small and medium-size enterprises are like grains of sand on a beach. The Internet can glue them together. It can make them into an invincible force that is able to go up against the big stones. The virtue of the Internet is that it uses small to compete with big, it uses fast to compete with slow.
>
> Jack Ma, the founder of e-commerce giant Alibaba (Liu and Avery 2009, p. 50)

Alibaba.com is the e-commerce giant, running B2B (business to business), B2C (business to customer) and C2C (customer to customer) e-marketplace that connects small and medium-size buyers and suppliers all over the world (Qing and Xue 2009). Jack Ma, a former English teacher, started Alibaba Company in 1999 with just $60,000. By 2014, Alibaba's profits surpassed those of Ebay and Amazon combined; only three technology companies in the world, i.e. Microsoft, Google and Apple, are bigger (Waite 2014). With Alibaba.com, barriers all around the world dissolve; anyone with Internet access has the opportunity to import, export, build commercial alliances, and trade with anyone else, anywhere in the world. In addition, customers can have a high degree of participation in the design of the products they purchase (Schepp and Schepp 2010). There is no doubt that the Alibaba.com platform can support VCIM implementation, in which component suppliers and assembly agents could be SMEs distributed worldwide. As a result, many SMEs, distributed locally and/or global, can get benefit from joining VCIM systems.

Production scheduling plays an important role in VCIM systems because it will affect the quality, cost and lead time of products. Quality of VCIM production scheduling depends on qualities of scheduling model and optimisation solution method. A comprehensive VCIM production scheduling model is the prerequisite for finding the global optimal solution or at least good solutions to the problem. Once having a comprehensive VCIM production scheduling model, quality of the scheduling will depend on the robustness of optimisation solution method. Developing a comprehensive VCIM production scheduling model and a robust optimisation solution method will make a significant contribution not only to VCIM systems but also to the larger fields such as Job Shop Scheduling, Vehicle Routing, Project Planning and Scheduling, Supply Chain Management, Computer Science, Mathematics, Engineering, Physics, Biochemistry, Genetics, Molecular Biology, etc., because it is possible to modify the developed scheduling model and GA to solve problems in other fields.

1.4 Research Objectives

This research has two objectives, i.e. developing a comprehensive VCIM production scheduling model and developing a robust optimisation solution method for the large-scale complex scheduling problem in the developed model. A comprehensive VCIM production scheduling model should be capable of (1) supporting collaborative shipment scheduling, (2) handling multiple product orders simultaneously, (3) dealing with multiple objective functions and (4) considering uncertainties. A robust optimisation solution method should be capable of searching for optimal/suboptimal solutions to the large-scale complex scheduling problem in the comprehensive VCIM production scheduling model, within a reasonable computing time.

1.5 Research Scope

This research only focuses on the production scheduling in VCIM systems producing mechanical products. VCIM production scheduling in this research is referred to as (1) selecting component suppliers, (2) selecting assembly agents and (3) scheduling the component and product shipments to fulfil the given product orders. How component suppliers, assembly agents, shipping providers and VCIM administrators communicate with each other, how to design VCIM products, how to decompose one product into a number of independent components, how component suppliers produce the components, how assembly agents assemble the products, and how shipping providers transport the components and products to the required destinations are outside the scope of this research. In other words, operational implementation of manufacturing, assembly and shipping functions in VCIM systems is not the subject of this research.

1.6 Thesis Structure

This thesis consists of six chapters. In this chapter, the introduction to the research, including research background, research motivation, research objectives and research scope, is presented. Chapter 2 is about the literature review on the development of Computer-Integrated Manufacturing, development of Virtual Enterprise, development of Virtual Computer-Integrated Manufacturing, production scheduling issue in VCIM systems, development of optimisation methods, development of Genetic Algorithms, and current research gaps. In Chap. 3, the developed VCIM production scheduling model, computational complexity and some discussions regarding the developed model are presented. Chapter 4 presents the developed GA for solving the scheduling problem in the developed VCIM production scheduling model which includes chromosome encoding, crossover, mutation, evaluation, selection, algorithm structure, parameter tuning and termination criterion. In Chap. 5, a comprehensive case study is given, which includes the problem description, data collection, implementation of the proposed approach, performance comparison, sensitivity analysis and some discussions. Finally, conclusions of the research, contributions of the research and recommendations for future work are provided in Chap. 6. For more details about the thesis structure, it is advised to see the Table of Contents.

1.7 Concluding Remarks

In this chapter, background of the research, i.e. production scheduling in VCIM systems, was presented. In addition, research motivation, which is to develop a foundation to build and operate VCIM systems that can help many small and medium-size manufacturing enterprises survive and thrive in today's competitive global market, was highlighted. Next, two objectives of the research, namely developing a comprehensive VCIM production scheduling model and developing a robust optimisation solution method for the large-scale complex scheduling problem in the developed model, were given. Finally, research scope and thesis structure were presented.

References

Angilella, S., & Mazzù, S. (2015). The financing of innovative SMEs: a multicriteria credit rating model. *European Journal of Operational Research, 244*(2), 540–554.

Camarinha-Matos, L. M., & Afsarmanesh, H. (1999). The virtual enterprise concept. In L. M. Camarinha-Matos & H. Afsarmanesh (Eds.), *Infrastructures for virtual enterprises—networking industrial enterprises* (pp. 3–14). The Netherlands: Kluwer Academic Publishers.

Costa, S. G., Platts, K., & Fleury, A. (2000). Advanced manufacturing technology: defining the object and positioning it as an element of manufacturing strategy. In *VI International Conference on Industrial Engineering and Operations Management*, (pp. 1–8).

Dao, S.D., Abhary, K., Marian, R., (2016a). A stochastic production scheduling model for VCIM systems, *Intelligent Industrial Systems, 2*(1), 85–101.

Dao, S. D., Abhary, K., & Marian, R. (2016b). Is a virtual computer-integrated manufacturing system feasible? In *Proceedings of the World Congress on Engineering*, (pp. 733–736). 29 June–1 July 2016, London, U.K.

Gunawardana, K. D. (2006). Introduction of advanced manufacturing technology: a literature review. *Sabaragamuwa University Journal, 6*(1), 116–134.

Iturrioz, C., Aragón, C., & Narvaiza, L. (2015). How to foster shared innovation within SMEs' networks: social capital and the role of intermediaries. *European Management Journal 33*(2), 104–115.

Koc, T., & Bozdag, E. (2009). The impact of AMT practices on firm performance in manufacturing SMEs. *Robotics and Computer-Integrated Manufacturing, 25*(2), 303–313.

Kumari, S., Singh, A., Mishra, N., & Garza-Reyes, J. A. (2015). A multi-agent architecture for outsourcing SMEs manufacturing supply chain. *Robotics and Computer-Integrated Manufacturing, 36*, 36–44.

Liu, S., & Avery, M. (2009). *Alibaba: the inside story behind Jack Ma and the creation of the world's biggest online marketplace*. New York: HarperCollins e-books.

Miller, F. P., Vandome, A. F., & McBrewster, J. (2010). *Computer-integrated manufacturing*. Mauritius: VDM Publishing House.

Nagalingam, S. V., & Lin, G. C. I. (1999). Latest developments in CIM. *Robotics and Computer-Integrated Manufacturing, 15*(6), 423–430.

Qing, H. H., & Xue, Z. S. (2009). A model for value-added e-marketplace provisioning: case study from Alibaba.com. In C. Godart et al. (Eds.), *Software Services for E-Business and E-Society* (Vol. 305, pp. 65–72). Berlin Heidelberg: Springer.

Schepp, B. & Schepp, D. (2010). *The official Alibaba.com success guide insider tips and strategies for sourcing products from the world's largest B2B marketplace*. Hoboken: Wiley.

Small, M. H., & Yasin, M. M. (1997). Advanced manufacturing technology: Implementation policy and performance. *Journal of Operations Management, 15*(4), 349–370.

Suresh, N. C., & Meredith, J. R. (1985). Justifying multimachine systems: an integrated strategic approach. *Journal of Manufacturing Systems, 4*(2), 117–134.

Waite, S. (2014). Record-breaking Alibaba allocates to happy few. *FinanceAsia (Deal of the month: October)* 8–9. http://ezlibproxy.unisa.edu.au/login?, http://search.ebscohost.com/login.aspx?direct=true&db=bth&AN=99088865&site=ehost-live.

Wang, D. (2007). *The development of an agent-based architecture for virtual CIM* (Ph.D. thesis). University of South Australia, Adelaide, Adelaide.

Yazdanifard, R., & Li, M. T. H. (2014). The review of Alibaba's online business marketing strategies which navigate them to present success. *Global Journal of Management and Business Research, 7*(9), 33–39.

Zhou, N., Xing, K., Nagalingam, SV., & Lin, G. (2010). Development of an agent based VCIM resource scheduling process for small and medium enterprises In *Proceedings of the International Multiconference of Engineers and Computer Scientists*, (pp. 39–44).

Chapter 2
Literature Review

2.1 Introduction

This research focuses on production scheduling in VCIM systems, i.e. developing a comprehensive production scheduling model for VCIM systems and developing a robust optimisation solution method for the complex scheduling problem in the developed model. In this chapter, a comprehensive critical literature review is presented, which includes the development of Computer-Integrated Manufacturing, development of Virtual Enterprise, development of Virtual Computer-Integrated Manufacturing, production scheduling issues in VCIM systems, development of optimisation methods, development of Genetic Algorithms, and current research gaps.

2.2 Computer-Integrated Manufacturing

Computer-Integrated Manufacturing (CIM), initially and conceptually coined by Harrington (1973), is a manufacturing system in which computers are used to control the production processes. All functional areas in the system such as design, planning, analysis, purchasing, inventory control, cost accounting, production, marketing, material handling, distribution, management, etc. are linked together by means of computer systems. Therefore, the manufacturing can be automated, faster and less error-prone (Miller et al. 2010). Effective integration of advanced manufacturing technologies such as Computer Aided Design (CAD), Computer Numerical Control (CNC), Computer Aided Process Planning (CAPP), Automated Material Handling Systems (AMHS), Flexible Manufacturing Systems (FMS), etc. in various functional units makes CIM systems powerful (Nagalingam and Lin 1999).

S. D. Dao, *Modelling and Intelligent Optimisation of Production Scheduling in VCIM Systems*, Springer Theses, https://doi.org/10.1007/978-3-319-72113-2_2

CIM systems are powerful but very expensive. It is very challenging to justify CIM investments because the benefits and costs associated CIM systems are difficult to quantify. To deal with the challenge, Nagalingam (1999) proposed a so called *innovative decision support system for CIM justification* in which a multi-objective approach, a cost value method to evaluate intangibles, a method to determine the effect of inter-functional benefits of different technologies and a goal programming resource allocation procedure were developed. This decision support system, a multi-attribute decision-making mechanism that accounts for both quantitative and qualitative factors, is capable of helping decision makers make a better CIM investment decision. Nevertheless, uncertainties or stochastic parameters were not taken into account there.

CIM systems are advanced manufacturing systems capable of producing the products with more complexity, higher degree of accuracy, more efficiency, shorter manufacturing lead-time, smaller inventory level and less cost, compared to the traditional manufacturing systems (Yurdakul 2004). However, CIM systems are very complex due to many advanced manufacturing technologies involved. Selecting CIM technologies that satisfy the constraints faced by the company is not a trivial task. To do so, a combined analytic hierarchy process and goal programming model was developed by Yurdakul (2004). Bozdağ et al. (2003) proposed the fuzzy group-decision making method for selecting the technologies to build a CIM system. In addition, an experimental design approach using TOPSIS method was proposed by Yusuf (2012) to select CIM technologies. Selecting the right technologies plays an important role in the success of CIM systems. However, research on optimisation of CIM technology selection is still very limited.

Once CIM systems have been established, operating them effectively is another challenge. A computer shop floor control model for VCIM systems was developed by Yang et al. (2000). Bal et al. (2008) developed a virtual-reality-based information requirement analysis tool for CIM system implementation. To measure the flexibility of CIM systems, a fuzzy cash flow analysis was proposed by Kahraman et al. (2004). To maximise the performance of CIM systems, Salehi and Moghaddam (2009) developed an innovative process planning method, in which the process planning was divided into preliminary planning and secondary planning, and genetic algorithm was used to search for optimal solutions to the problem. Due to advances in computer science and information technology, CIM systems operate very effectively and are widely used in manufacturing industries (Yu et al. 2015). Nevertheless, a comprehensive research on optimisation associated with CIM system operation is still missing.

CIM systems are much more capable than the traditional manufacturing systems, due to the effective integration of various advanced manufacturing technologies (Nagalingam and Lin 1999). Nevertheless, CIM systems can only exploit the local manufacturing resources and their capability will be limited if the resources are not always available (Wang 2007). To overcome this limitation of CIM systems, more flexible manufacturing systems, i.e. Virtual Enterprise and Virtual Computer-Integrated Manufacturing, have been being developed to help small and medium-size manufacturing enterprises survive and thrive in today's competitive global market. The developments of Virtual Enterprise and Virtual Computer-Integrated Manufacturing will be presented in the following Sections.

2.3 Virtual Enterprise

There has been a significant amount of research focused on Virtual Enterprise since this concept emerged in the early 1990s. There are a number of definitions for Virtual Enterprise concept but the most popular one is as follows: "Virtual Enterprise is a temporary alliance of enterprises that come together to share skills, core competencies and resources in order to better respond to business opportunities" (Camarinha-Matos and Afsarmanesh 1999, p. 4).

Virtual Enterprise is capable of effectively exploiting the competitive advantages of the member enterprises by temporary cooperation and networking. The effective cooperation enables Virtual Enterprise to produce the products with higher quality, lower cost, and shorter lead time (Huang et al. 2002). In addition, collaborative manufacturing is a clear trend nowadays; each company is just one node adding some value to the entire production cycle (Camarinha-Matos and Afsarmanesh 1999). Clearly, temporary collaboration is one of the unique characteristics of Virtual Enterprise. Concept of Virtual Enterprise has become a reality and it is a powerful business solution to many companies worldwide in today's competitive global market (Castro et al. 2013; Mun et al. 2009).

Virtual Enterprise is generally formed as follows. There is an enterprise having a big project in hand. However, this enterprise, called principal enterprise, does not complete the project alone because of several reasons such as limited capability, lack of resources or even just wishing to better use its own core business competence. Therefore, the principal enterprise breaks down the project into a number of sub-projects and selects some of those sub-projects to be completed by itself. And then, the principal enterprise invites other enterprises to tender the rest of the sub-projects. A number of appropriate partner enterprises are selected to complete the sub-projects through bidding. As a result, the co-operation between the principal enterprise and its partners forms Virtual Enterprise that can complete the project on time, with a higher quality standard and lower cost. When the project is finished, this co-operation is dissolved (Crispim and Sousa 2009; Huang et al. 2011; Huang and Fan 2007; Ip et al. 2003; Simona and Raluca 2011; Tao et al. 2010; Wu and Su 2005; Ye 2010; Ye and Li 2009; Zeng et al. 2006; Zhang et al. 2012; Zhao et al. 2008).

Partner selection is a critical issue in the success of Virtual Enterprise. Typically, there are four stages in the life cycle of Virtual Enterprise: *creation*, *operation*, *evolution* and *dissolution* (Wu and Su 2005; Ye 2010; Zhong et al. 2009). The success of Virtual Enterprise depends heavily on the creation stage in which the right partner enterprises should be selected. The partner selection problem in Virtual Enterprise has been attracting a lot of research attention. However, still a lot of works needs to be done to find a comprehensive optimisation solution method for the problem.

In the research by Niu et al. (2012), an Enhanced Ant Colony Optimiser for multi-attribute partner selection in Virtual Enterprise was developed. The candidate partners were evaluated based on both qualitative and quantitative aspects, i.e. cost, time, quality, reputation and risk. Experimental data showed that the Enhanced Ant Colony Optimiser could provide better results in terms of search accuracy and computing time, compared to the original Ant Colony Optimiser. Nevertheless, this selection model is not very comprehensive because of not accommodating uncertainties. In addition, much more benchmark algorithms are required to thoroughly verify the effectiveness of the developed Enhanced Ant Colony Optimiser.

Zhang et al. (2013) developed a green model for partner selection in Virtual Enterprise, in which two green criteria, namely carbon emission and lead content, were considered. To solve the problem in this model, Pareto Genetic Algorithm was developed. With Pareto idea, vector encoding, random selection, two-point crossover and single-point mutation, the Pareto Genetic Algorithm outperformed the two other optimisation algorithms, i.e. Simulated Annealing and Particle Swarm Optimisation. However, uncertainties or stochastic parameters in the green partner selection model were not taken into account yet.

Dao et al. (2014) developed an innovative Genetic Algorithm with unique chromosome representation and modified genetic operators, for partner selection and collaborative transportation scheduling in Virtual Enterprise. Effectiveness of the proposed approach was demonstrated through a comprehensive case study. Again, this is a deterministic partner selection model only since the stochastic parameters were not taken into consideration.

Recently, a stochastic partner selection approach for Virtual Enterprise was proposed by Crispim et al. (2015) where both stochastic and deterministic criteria for the partner selection were taken into account. Chance constraints were used to incorporate stochastic features into the multi-objective partner selection model. Alternative configurations of Virtual Enterprise in business environments with uncertainties were ranked by the chance-constrained method. In addition, multi-objective Tabu Search combined with a 2-Tuple Linguistic Procedure was used to support the selection. The preliminary computational results revealed the effectiveness of the proposed approach. Nevertheless, the optimisation solution method to solve the problem still could be better if more advanced optimisation algorithms like Particle Swarm Optimisation, Cuckoo Search, or Genetic Algorithm are used.

Beside the partner selection, there are a number of research topics in Virtual Enterprise such as enabling technologies, supporting platforms, information infrastructures, and Virtual Enterprise controls (Huang et al. 2002). Significant developments in Virtual Enterprise implementation have been achieved so far. However, the capability of Virtual Enterprise is limited due to boundaries still existed between different enterprises. Based on the developments of Virtual Enterprise and Computer-Integrated Manufacturing, it is believed that these two systems are moving to the next stage, namely Virtual Computer-Integrated Manufacturing (Nagalingam et al. 2007; Wang 2007; Wang et al. 2007). The latest development of Virtual Computer-Integrated Manufacturing will be reviewed in the next Section.

2.4 Virtual Computer-Integrated Manufacturing

Concept of Virtual Computer-Integrated Manufacturing (VCIM) was first introduced in a keynote speech (Lin 1997) in the 4th International Conference on Computer-Integrated Manufacturing in Singapore. In the keynote speech, the latest research trends in Computer-Integrated Manufacturing (CIM) were discussed; and then based on the observation of global market condition and rapid technology development at that time, the author predicted that different CIM systems would be integrated together and this integration would go beyond geographical boundaries. This prediction was a starting point of VCIM concept.

The term, VCIM, was first used by Nagalingam and Lin (1999). It should be noted that the letter "V" in VCIM stands for "Virtual", which was borrowed from the concept of Virtual Enterprise. As mentioned in Sect. 2.3, the concept of Virtual Enterprise has become a reality and it is a powerful business solution to many companies worldwide in today's competitive global market. Based on the developments of CIM systems, technologies, Virtual Enterprise and global market, Nagalingam and Lin (1999) suggested that research should be done to develop a global, flexible and integrated manufacturing system called VCIM. However, the definition of VCIM was not very clear then.

Nagalingam and Lin (2000) defined VCIM as a network of interconnected global CIM systems. This definition has two important characteristics. First, VCIM is an integrated manufacturing system in which all of its units are connected to one another by means of computer networks. Second, VCIM is a global manufacturing system in which its units are distributed locally and/or globally. According to them, building and running a VCIM system require a number of things such as a collaborative framework for remote machining, a communication protocol, an integrated architecture, Internet technology, centralised and localised databases, an autonomous decision-making mechanism, etc. Although the global integration was clearly expressed in the definition of VCIM, the flexibility like the one in Virtual Enterprise was not sufficiently emphasised then.

Wang et al. (2003b) proposed the agent-based approach to study and implement VCIM where argued that the agent-based approach is a suitable approach to study and implement VCIM because VCIM is a distributed and integrated manufacturing system. Agents themselves and their interactions with each other could represent a VCIM system well. The agent-based VCIM model had six types of agents, i.e. administration agent, design agent, engineering service agent, manufacturing agent, marketing agent, and financial agent. All of those agents were connected to one another via the Internet infrastructure. As can be seen, the agent-based VCIM model was quite comprehensive. Nevertheless, how the VCIM system works was not described in their work as a consequence of which they did not evaluate the performance of the agent-based VCIM model.

An implementing architecture for VCIM was discussed by Nagalingam et al. (2003) where they highlighted a number of issues needed to be solved to implement a VCIM system. First, real-time connections among agents of a VCIM system must be established. Second, all agents must have a consistent data format. Third, object-oriented technology and common object request broker architecture are required to bring the different distributed agents together to complete the customer orders. Fourth, a common database must be set up to manage the facilities and resources effectively. However, their VCIM model was still not comprehensive since uncertainties, multiple objective functions and multiple product orders were taken into consideration.

In a later research, Wang et al. (2004) developed a new architecture called a parallel processing multi-agent architecture for VCIM systems which was capable of providing multiple coordinator agents with similar functionality, so that the agents could coordinate the information flow across a VCIM system in a parallel manner. A simulation demonstrated that VCIM systems with the parallel processing multi-agent architecture could operate much more efficiently, compared to systems with the then existing architectures. Nevertheless, much more comprehensive and realistic case studies are required to thoroughly verify the robustness of the parallel processing multi-agent architecture.

In a Ph.D. thesis on developing an agent-based architecture for VCIM systems by Wang (2007), the author argued that VCIM systems would help many SMEs survive and thrive in the highly competitive global market, because VCIM systems are very flexible and capable of exploiting locally and/or globally distributed resources. In a VCIM system, manufacturing resources may belong to different partner enterprises or may be located at different regions; however all partner enterprises are willing to work together in an integrated manner to achieve market share. According to this author, to implement a VCIM system, a number of issues need to be solved such as how to interconnect distributed resources to form an integrated manufacturing system, how to dynamically schedule and organise distributed manufacturing resources as a temporary production system to produce a given product, and how to route information and material flow across a VCIM system. To support VCIM implementation, an agent-based architecture was developed by this author, in which there were three types of agents, namely facilitator agent, customer agent and resource agent. The facilitator agent acts as a coordinator to manage information flow across a VCIM system. The customer agent is to provide graphical user interface for customers to use VCIM services. The resource agent represents the manufacturing functionality in a VCIM system. In addition, the facilitator agent in this architecture is capable of (1) processing product order in a parallel manner and (2) incorporating an algorithm to optimise the resource allocation in a VCIM system. Furthermore, a prototype of the agent-based VCIM architecture was built and some insights of the system were reported. However, performance of the proposed approach was not compared to others; in other words, there was no performance benchmark in that research.

Zhou et al. (2010a, b) developed an agent-based VCIM resource scheduling model for small and medium-size manufacturing enterprises. This model aims to unite and integrate all activities in a network of small and medium-size manufacturing enterprises to share the resources to better respond to the constantly changing business opportunities in today's global market. In this model, manufacturing resources may belong to different partner enterprises or may be located at different regions; however all partner enterprises are willing to work together in an integrated manner as a temporary manufacturing system to fulfil a given customer order. When the order is fulfilled, the temporary manufacturing system will vanish. This temporary cooperation between different manufacturing enterprises makes VCIM systems very flexible in exploiting the distributed resources. It should be noted that this characteristic, namely temporary cooperation, was originated from the concept of Virtual Enterprise. Although the agent-based VCIM resource scheduling model was described in detail, a case study was not conducted there to verify the effectiveness of the model.

The above literature review indicates that VCIM inherits two unique characteristics, i.e. integration and temporary cooperation, from two mature concepts, Computer-Integrated Manufacturing and Virtual Enterprise, respectively. VCIM is a new global manufacturing system, which is very integrated and flexible. In addition, majority of the published research focused on developing the agent-based VCIM models; and several prototypes of VCIM models have been built, though are not comprehensive.

2.5 Production Scheduling in VCIM Systems

Production scheduling is very important in VCIM systems because it will affect the quality, cost, and lead time of products (Dao et al. 2016a). After receiving product orders, the VCIM system decomposes the product orders into a number of components which can be independently produced; and then it does the production scheduling by (1) selecting some component suppliers to produce the required components, (2) selecting some assembly agents to assemble the required products and (3) scheduling the shipments to transport the components and products to the required destinations, so that a temporary VCIM production system can be formed to fulfil the product orders. The whole procedure will be repeated for other product orders. VCIM production scheduling is a very complex problem, which requires a comprehensive production scheduling model and a robust optimisation solution method.

Production scheduling in VCIM systems is not exactly the same as production and/or resource scheduling in various traditional manufacturing systems like flexible robotic manufacturing cells (Batur et al. 2012), flexible robotic assembly cells (Abd et al. 2016), flexible manufacturing system (Cardin et al. 2013), make-to-order manufacturing system (He et al. 2014; Sawik 2007), etc. There are a number of differences between VCIM production scheduling and the traditional production scheduling. First, VCIM production scheduling is an integrated optimisation problem since it has a number of interlinked sub-problems of different natures, namely manufacturing agent selection, assembly agent selection, component shipment scheduling and product shipment scheduling. Second, these sub-problems should be simultaneously solved otherwise the obtained solutions may not be feasible. Third, VCIM production scheduling is a dynamic optimisation problem because the number of manufacturing and assembly agents selected in each solution is variable.

Nagalingam and Lin (2000) developed a VCIM production scheduling model as follows. When customer orders a product, the information will go to the central database. The autonomous decision-making mechanism of the VCIM system is then activated to find the best location (a locally controlled manufacturing system) to produce the requested product. During this decision making process, data is exchanged between the central and local databases. After the final decision is confirmed, the selected locally controlled manufacturing system will produce the product and then ship it to the customer. The whole procedure will be repeated for the next customer order. However they did not mention how the autonomous decision-making mechanism finds the best locally controlled manufacturing system to produce the requested product. This implies that this VCIM production scheduling procedure is too general.

Wang et al. (2003a) developed an agent-based VCIM model in which the production scheduling was done as follows. When receiving a product order, the facilitator agent decomposes the order into a number of sub-tasks and then sends them to the available resource agents. The resource agents analyse the received sub-tasks by using their built-in expert systems before responding to the facilitator agent. After receiving the responses, the facilitator agent selects the most appropriate resource agents to complete the sub-tasks and send them the confirmation messages. Finally, the production schedule is built to fulfil the product order. Obviously, this production scheduling is an optimisation problem as many feasible solutions for producing one product may exist. Selecting the best solution among many feasible ones has never been a trivial task especially for complex large-scale problems. However, such selection, i.e. selecting the most appropriate resource agents to complete the sub-tasks was overlooked here.

In a later research, Wang et al. (2004) proposed the following VCIM production scheduling procedure. The facilitator agent decomposes the received product order into a number of interrelated sub-tasks. For each sub-task, the facilitator agent selects a number of potential resource agents and sends them a request message. When receiving the request to complete the sub-task, if capable, the resource agent will generate a production proposal which contains the cost and completion time. After receiving all production proposals for the sub-task from the selected resource agents, the facilitator agent select the best production schedule for the sub-task, based on the cost and completion time. When the best production schedules for all sub-tasks are selected, the full production schedule for the product order is compiled. Again, the selection procedure here was overlooked. In addition, adding the best production schedules for the sub-tasks may not result in the best full production schedule for the product.

The VCIM production scheduling procedures developed by a group of researchers (Nagalingam et al. 2007; Wang 2007; Wang et al. 2007; Zhou et al. 2010a, b) can be summarised as follows:

Step 1	Decompose the received product orders into a number of parts
Step 2	Select manufacturing agents to produce the decomposed parts in Step 1
Step 3	Select transportation agents to transport the decomposed parts directly from their manufacturing agents to the required assembly agent(s)
Step 4	Form full schedules using decisions in Steps 2–3
Step 5	Find the best full schedule among those in Step 4

As can be seen from Steps 1–5, the above model could provide decision makers with feasible solutions to the VCIM production scheduling problem. However, this scheduling model is not comprehensive because it does not consider collaborative transportation in spite of the fact that VCIM systems have a great potential for collaborative transportation since all agents in a VCIM system are willing to work together in an integrated manner. In addition, this scheduling model is not fully integrated because the full schedules in Step 4 are formed by just combining two fixed sub-schedules in Steps 2–3. Without a fully integrated scheduling model, it is impossible to achieve the global optimal solution.

Backward Network Algorithm was developed by Wang et al. (2007) to select the best VCIM production schedule. The developed Backward Network Algorithm is quite simple to implement and it is capable of handling large-scale problems; nevertheless, its capability of obtaining the global optimal solution is very limited, due to the way it constructs the full schedule from the fixed sub-schedules. Also a Genetic Algorithm was developed by Dao et al. (2012) to optimise the production scheduling in VCIM systems. Nevertheless, the global optimal solution to the VCIM production scheduling problem was yet to be achieved as the production scheduling model was not comprehensive because a number of issues such as multiple product orders, multiple objectives, uncertainties and collaborative transportation were not taken into account.

Transportation cost is a significant part of the total cost of the products because VCIM is a global manufacturing system. To reduce transportation cost, a number of researchers and practitioners have been developing collaborative transportation strategy where two or more carriers or shippers form partnerships to share vehicle capacities as well as delivery tasks (Asawasakulsorn 2009; Chan and Zhang 2011; Dai and Chen 2009; Ergun et al. 2007; Li and Chan 2012; Ozener 2008). It can be seen that collaborative transportation strategy has great a potential application in VCIM systems because all of the resource agents in a VCIM system are willing to work together in a fully integrated manner. However, collaborative transportation scheduling has not been considered in any existing VCIM production scheduling model yet.

As can be seen from the above literature review, currently there are two major limitations in VCIM production scheduling. First, the existing VCIM production scheduling models are not comprehensive because the collaborative shipment scheduling, multiple objectives, multiple product orders, uncertainties are not taken into consideration. Second, the existing optimisation methods for solving the VCIM production scheduling problem are not very robust since all sub-problems are not simultaneously solved. Without simultaneous solution of all sub-problems, it is impossible to achieve the global optimal solution.

2.6 Optimisation Methods

Optimisation is generally referred to as finding the best solution to an optimisation problem. Optimisation is always desirable in many fields such as engineering design, computer science, operations research, economics, computational chemistry, biomedicine, etc. (Coelho et al. 2014; Ng and Li 2014; Wang et al. 2013b). Optimisation methods can be broadly classified into two categories: deterministic methods and stochastic methods (Hanagandi and Nikolaou 1998). Each category has advantages and disadvantages. Deterministic methods are capable of guaranteeing the optimality for certain problems thanks to exploiting some helpful features of the problem structure. However, they may fail when tackling black-box problems, extremely ill-behaved functions or complex large-scale problems, due to the combinatorial explosion issue. Stochastic methods can work with any kind of optimisation problem but they are of weak capability of guaranteeing the optimal solutions. Stochastic methods only provide the optimal solutions with probabilistic guarantee and this probability tends to unity in infinite computing time (Liberti and Kucherenko 2005; Moles et al. 2003). Nevertheless, there is no algorithm capable of solving general optimisation problem with certainty in finite computing time (Boender and Romeijn 1995). These two optimisation solution approaches will be explored further below.

2.6.1 Deterministic Optimisation Methods

Deterministic optimisation methods are capable of guaranteeing the optimal solutions for certain problems thanks to exploiting some helpful features of the problem structure. There are a number of deterministic optimisation methods such as Branch-and-Bound methods, Cutting Plane methods, Primal-Dual Decomposition methods, Outer Approximation methods, Inner Approximation methods, Difference of Convex methods, Reverse Convex methods, Reformulation-Linearization methods, Lipschitzian methods, Trajectory and Homotopy methods, Interval Analysis methods, etc. (Floudas 2000).

Hfaiedh et al. (2015) applied Branch-and-Bound method to solve a single machine scheduling problem with release dates and unavailable machines during a fixed interval. The objective function in the scheduling problem is minimising the maximum delivery time. From the computational complexity theory point of view, this scheduling problem is an NP-hard optimisation problem. In the Branch-and-Bound method, Jackson's pre-emptive algorithm with precedence constraints was used to calculate the lower bound and Schrage's sequence was used to determine the upper bound. The effectiveness of the developed Branch-and-Bound method was verified in a number of numerical examples. Nevertheless, solving complex large-scale optimisation problems with the developed algorithm is not practical because a massive computing time would be required.

Fang et al. (2015) developed an exact approach based on the Cut-and-Solve and Cutting Plane methods for solving the capacitated lane reservation problem with residual capacity. The objective here is to minimise the impact of lane reservation. In this approach, a number of new algorithms were developed to find appropriate valid inequalities to speed up the optimal convergence of the proposed approach. Experimental data showed that the proposed approach outperforms other algorithms and CPLEX solver in the literature. However, a massive computing time would prevent the developed algorithm from solving complex large-scale problems.

Ackooij and Malick (2016) proposed a Primal-Dual Decomposition approach to solve the unit-commitment problem in electricity generation companies. In this research, the two-stage formulation of the unit-commitment problem was first proposed and the Primal-Dual Decomposition approach was then applied to solve it. In the proposed approach, the warm-started bundle algorithms were used and specific knowledge of underlying technical constraints was not required. Performance of the proposed Primal-Dual Decomposition approach was tested in a number of unit-commitment instances. Nevertheless, this is a deterministic problem only since uncertainties were not taken into consideration.

An Outer Approximation method for an integrated problem, called supply chain network design and assembly line balancing with demand uncertainty, was developed by Yolmeh and Salehi (2015) to solve the integrated problem formulated as a mixed integer nonlinear programming. The effectiveness of the developed Outer Approximation method was demonstrated by experimental data. However, multiple objective functions were not considered in their work.

Wozabal (2012) applied the Difference of Convex method to solve the Value-at-Risk constrained Markowitz style portfolio selection problem in financial institutions. This is a non-convex stochastic optimisation problem. A Difference of Convex program was used to reformulate the problem and then a Difference of Convex algorithm was developed to solve it. The robustness of the developed Difference of Convex method was demonstrated in a number of numerical examples. Nevertheless, this is a single objective optimisation method only.

Lan et al. (2016) developed a Branch-and-Reduce method based on Reformulation Linearization technique to solve the regional water supply system planning problem. In this research, a non-convex nonlinear model was formulated. A three-pronged effort associated with the space transformation of decision variables, polyhedral outer approximations and Reformulation Linearisation was used to construct the lower bound. The convergence was accelerated by the Range Reduction method. Efficiency of the proposed approach was verified by numerical results. Again, this planning model is not comprehensive because uncertainty and multi-objective optimisation were taken into consideration.

As can be seen from the above literature review, a number of deterministic optimisation methods have been developed to solve the optimisation problems in various fields such as Supply Chain, Finance, Transportation, Machine Scheduling, etc. which are capable of guaranteeing the global optimal solutions for certain problems thanks to exploiting some helpful features of the problem structure. Nevertheless, deterministic optimisation methods may fail when dealing with black-box problems, extremely ill-behaved functions or complex large-scale problems (Dao et al. 2016b).

2.6.2 *Stochastic Optimisation Methods*

Stochastic optimisation methods can work with any kind of problem but they have a weak capability of guaranteeing the global optimal solution. Stochastic approaches provide the global optimal solution with probabilistic guarantee only and this probability will become 1 in infinite computing time (Liberti and Kucherenko 2005; Moles et al. 2003). Nevertheless, there is no algorithm capable of finding the global optimal solution with certainty to a general optimisation problem in finite computing time (Boender and Romeijn 1995).

Literature survey conducted in this research shows that stochastic optimisation methods are more popular than deterministic ones in real-life applications. This could be due to a number of reasons. First, stochastic approaches do not require sophisticated mathematical analysis to solve problems. Second, stochastic approaches could handle practical and large-scale problems better than deterministic ones. Finally, nowadays the advanced computational technology allows stochastic approaches to increase the probability of finding the global optimal solution because a large number of solutions can be generated and evaluated in a relatively short computing time (Dao et al. 2016b).

A number of stochastic optimisation methods have been developed to solve many complex large-scale problems such as Tabu Search, Particle Swarm Optimisation, Cuckoo Search, Ant Colony Optimisation, Simulated Annealing, Hill Climbing, Downhill Simplex, Artificial Bee Colony Algorithm, Swarm Intelligence, Differential Evolution Algorithm, Genetic Algorithm, etc. Each method has advantages and disadvantages.

Shahvari and Logendran (2017) developed a Tabu Search algorithm to solve a sequence- and machine-dependent batch scheduling problem in the environment of unrelated-parallel machines. The objective function here is minimising a linear combination of completion time and tardiness. To solve the problem they first developed a mixed integer linear program model, and then a Tabu Search algorithm with multi-level diversification and multi-tabu structure. Experimental data revealed that the developed Tabu Search algorithm outperforms the commercial optimisation software package CPLEX. Nevertheless, this scheduling model is not comprehensive either because it did not include uncertainties and multiple objective functions.

Lin et al. (2016) developed a Particle Swarm Optimisation algorithm for the two-agent multi-facility order scheduling with ready times. In this problem, the total completion time of the orders of one agent was the objective function to be minimised. A Branch-and-Bound algorithm was first proposed to incorporate a number of dominance rules and a lower bound to the problem, and then their Particle Swarm Optimisation algorithm was used to search for the optimal/near-optimal solutions to the problem. In the developed algorithm, a number of techniques such as two-level particle number, neighbourhood improvement, fixed inertia weight and decreasing inertia weight were applied. Efficiency of the proposed Particle Swarm Optimisation algorithm was demonstrated in many case studies.

Majumder and Laha (2016) developed a Discrete Cuckoo Search algorithm to solve a scheduling problem in two-machine robotic cells. The scheduling question here is to determine the sequences of robots and parts in a robotic cell so that the cycle time will be minimised. In the developed algorithm, a fractional scaling factor based method was used to determine the step length of Levy flight distribution. In addition, two neighbourhood search methods, namely Interchange and Cyclical Shift methods, were applied to improve the solution quality. Moreover, a Response Surface Methodology was used to accelerate the convergence of the algorithm. Furthermore, the parameters of the developed algorithm were systematically tuned by Design of Experiment. Effectiveness of the developed Discrete Cuckoo Search algorithm was thoroughly verified in a large number of numerical examples.

Wan et al. (2016) developed a modified Ant Colony Optimisation algorithm combined with Genetic Algorithm to solve the feature selection problem in data mining and pattern recognition. The purpose here is to select the feature subset to maximise the discriminating ability and minimise the redundancy. In this method, there were two models, namely the visibility density model and the pheromone density model. The solution found in the visibility density model was used as the visibility information, while the one in the pheromone density model was used as the initial pheromone information. Robustness of the developed algorithm was compared with a number of existing algorithms in the literature.

Akram et al. (2016) developed a Hybrid Simulated Annealing algorithm to solve the job shop scheduling problem. In the developed algorithm, a fast Simulated Annealing was used for the global search while Quenching technique was used for the local search. In addition, Tabu list was used to avoid revisiting the previously explored solutions. Effectiveness of the developed Hybrid Simulated Annealing algorithm was thoroughly verified in 88 well-known benchmark problems in the literature.

Gao et al. (2016) proposed an Improved Artificial Bee Colony algorithm to deal with the flexible job-shop scheduling problem with fuzzy processing time. Two objective functions, namely minimising the maximum fuzzy completion time and maximising the fuzzy machine workload were considered in this scheduling problem. In the proposed approach, an effective heuristic rule was used to generate the initial population. Robustness of the Improved Artificial Bee Colony algorithm was demonstrated in a number of benchmark problems as well as real-life case studies in remanufacturing industry.

As can be seen from the above literature review, many stochastic optimisation methods have been developed so far and many complex practical optimisation problems in various fields have been solved by the methods. Every algorithm has its own particularities, strengths and weaknesses. Nevertheless, Genetic Algorithm is one of the most popular stochastic optimisation methods, often used to deal with complex large-scale optimisation problems (Marian et al. 2012; Shahlaei et al. 2012). Genetic Algorithm has a number of advantageous characteristics, compared to other stochastic optimisation methods, such as flexibility in defining constraints as well as quality measures, capability of working with both continuous and discrete variables, capability of handling large search space, capability of providing multiple optimal/good solutions, and great potential for applying parallel computing techniques to shorten the processing time (Fahimnia et al. 2008). That is why in this research Genetic Algorithm based approach is chosen to solve the VCIM production scheduling problem. A comprehensive literature review on Genetic Algorithm will be presented in the next Section.

2.7 Genetic Algorithms

Genetic Algorithm (GA) is a popular solution method, often used to optimise solutions to problems in many fields such as engineering, computer science, economic management, supply chain management, etc. (Aguilar-Rivera et al. 2015; Cheng and Chang 2007; Lee et al. 2012; Su et al. 2015). As mentioned in Sect. 2.6, GA has a number of advantageous characteristics compared to other optimisation methods and it is the most suitable tool to solve the VCIM production scheduling problem.

Since being introduced over four decades ago, significant developments in various components of GA such as chromosome encoding (Dao et al. 2014; Dao and Marian 2011c; Zhong and Chen 2002), crossover (Qing-dao-er-ji and Wang 2012; Wang and Zheng 2002), mutation (Tang and Tseng 2013; Wang et al. 2009), evaluation (Chang et al. 2007; Hyun et al. 1998), selection (Stern et al. 2006; Wu et al. 2007) as well as algorithm structure (Dao et al. 2015; Zhou et al. 2011) have been made. These developments make today's GA much more powerful than the traditional GA. It should be noted that GA is a searching philosophy only; there is no standard GA. Generally speaking, when using GA, users must customise some GA components, e.g. chromosome encoding, crossover or mutation. Due to the increasing computing power, application of GA has been expanded to many different fields. It would be interesting to see a big picture of the developments of GA throughout its history. The bibliometrics of GA is presented in the next Section.

2.7.1 *Bibliometrics of Genetic Algorithms*

To show a big picture of the developments of GA throughout its history, a bibliometric analysis is conducted herein using Scopus database. With the search engine "*Title, Abstract, Keywords*" and with the keywords "Genetic Algorithm" OR "Genetic Algorithms", 124,799 publications associated with GA published from the beginning of GA history to 2014 were found (accessed on 12 November 2015). The detail of this bibliometric analysis has been published in a journal article (Dao et al. 2017). Some key points from that bibliometric analysis are summarised as follows. Majority of publications associated with GA are in the fields of engineering, computer science and mathematics. Most of the publications are in forms of journal articles and conference papers. The first documents associated with GA were published in 1972. The quantity of the publications has been significantly increasing since 1992 with the average increase rate of 475.7 publications per year.

2.7.2 Latest Developments of Genetic Algorithms

Genetic Algorithm (GA), first introduced by Holland (1975) is a powerful stochastic search algorithm based on the mechanisms of natural genetics and selection (Goldberg 1989). A general description of GA is as follows: "*Genetic algorithm ... starts with an initial set of random solutions called population. Each individual in the population is called a chromosome representing a solution to the problem at hand. The chromosomes evolve through successive iterations, called generations. During each generation, the chromosomes are evaluated using some measures of fitness. To create the next generation, new chromosome, called off-spring, are formed by either (a) merging two chromosomes from current generation using a crossover operator or (b) modifying a chromosome using a mutation operator. A new generation is formed by (a) selecting, according to the fitness values, some of the parents and offspring and (b) rejecting others so as to keep the population size constant. Fitter chromosomes have higher probabilities of being selected. After several generations, the algorithms converge to the best chromosome, which hopefully represents the optimum or suboptimal solution to the problem*" (Gen and Cheng 1997, pp. 1–2).

The traditional structure of GA is shown in Fig. 2.1. It can be seen that a typical GA has five main components: chromosome encoding, crossover, mutation, evaluation and selection. There have been significant developments in those components such as binary encoding (He and Hui 2010), real encoding (Deep et al. 2009), integer-number encoding with fixed length (Qu et al. 2013), hybrid encoding (Chen et al. 2006), variable length chromosome encoding (Dao and Marian 2011b), multi-dimensional chromosome encoding (Dao et al. 2014), single-point crossover (Dao and Marian 2011a), multi-point crossover (Esen and Koç 2015), heuristic crossover (Balakrishnan et al. 2003), comparison crossover (Maity et al. 2015), hybrid crossover (Suresh et al. 2014), adaptive directed mutation (Tang and Tseng 2013), power mutation (Deep and Thakur 2007), uniform/non-uniform mutation (Michalewicz 1996), single objective evaluation (Faghihi et al. 2014), multi-objective evaluation (Aiello et al. 2013), tournament selection (Castelli and

Vanneschi 2014), roulette-wheel selection (Zhao and Wang 2011), probabilistic selection (Maity et al. 2015), dynamic ranking selection (Boudissa and Bounekhla 2012), etc. Chromosome encoding, crossover, mutation, evaluation as well as selection methods have a certain effect on performance of GA and choosing them is dependent on property of problems under consideration.

Beside the five main components mentioned above, performance of GA is also affected by the algorithm structure. To enhance the search capability of GA, many researchers have attempted to improve the traditional structure of GA as shown in Fig. 2.1 by integrating local search techniques such as Simulated Annealing, Tabu Search, Hill Climbing, Pattern Search, Ant Colony Optimisation, Particle Swarm Optimisation, etc. into GA to create new algorithm called hybrid GA (Akpınar et al. 2013; Mahmoodabadi et al. 2013; Shokouhifar and Jalali 2015a, b; Wang et al. 2013a; Yun et al. 2013). Generally speaking, hybrid GA has better exploitation which is a capability to exploit some promising regions of the search space, compared to traditional GA. However, capability of hybrid GA to explore the search space, so called exploration, is still limited because solutions obtained by hybrid GA are evolved from only one set of solutions, usually, randomly generated in the first generation.

One way to improve the exploration of GA is to increase its population size. Nevertheless, too large population size would make the searching process of GA less efficient. Another way is *multistart* procedure which is simply to run GA for a number of times and the best solution among those achieved in the runs can be considered as the global optimal solution if the number of runs is large enough (Dao et al. 2014; Dao and Marian 2013). However, the existing *multistart* procedure is less capable of exploiting the elite chromosomes from each run as well as balancing between the exploration and exploitation.

Regarding the latest applications of GA in solving the scheduling and/or planning problems, Kundakcı and Kulak (2016) applied GA to solve the dynamic Job Shop scheduling problem. There were a number of dynamic events in the problem such as machine breakdowns, random job arrivals, changes in processing time, etc. Objective function here was to minimise the makespan. To solve the problem, an efficient hybrid GA was developed. Effectiveness of the developed GA was demonstrated in a number of benchmark problems.

Li et al. (2016) developed a Branch Population GA to solve the Dual Resource Constrained Job Shop Scheduling problem. There were two objective functions to be minimised here, i.e. cost and makespan. In the developed GA, the branch population was used to accumulate and transfer the evolutionary experience of parent chromosomes as a result of which population diversity and convergence of the algorithm was enhanced. Two other techniques, i.e. elite evolutionary operator and roulette selection operator based on sector segmentation, were also applied to improve the solution quality. Robustness of the developed Branch Population GA was verified in various numerical examples and case studies.

Xia et al. (2016) developed a hybrid GA with variable neighborhood search to solve the dynamic integrated process planning and scheduling problem. Machine

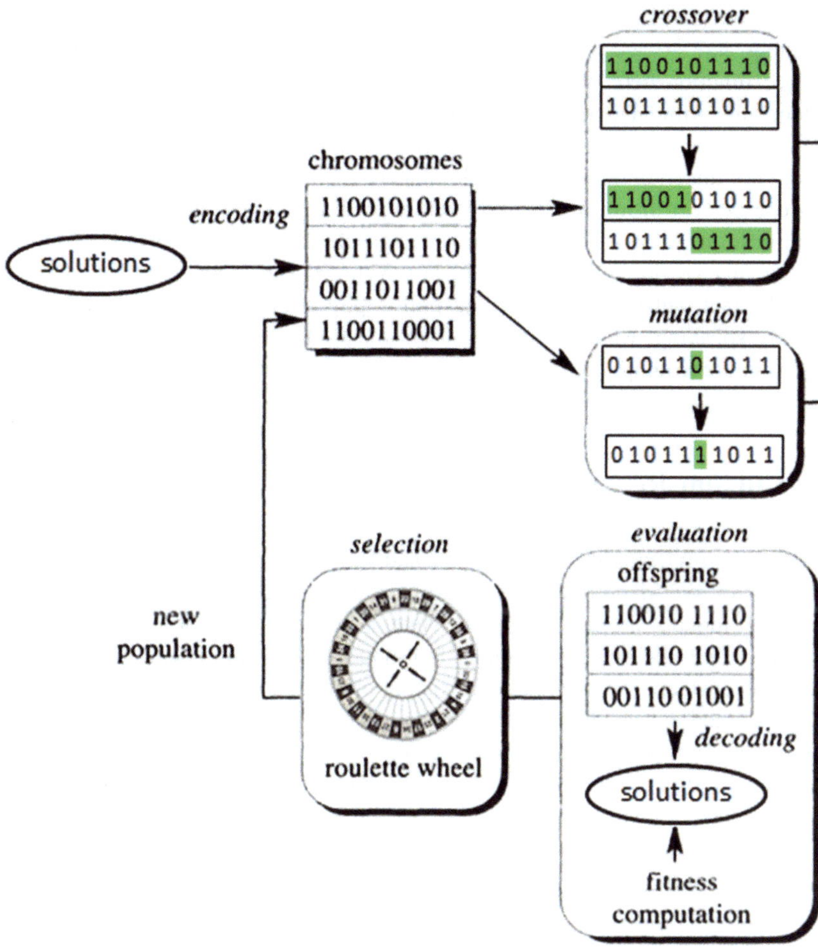

Fig. 2.1 Traditional structure of GA. Adapted from Gen and Cheng (1997, p. 3)

breakdown and new job arrival were two sources of uncertainties in this problem. Through three case studies, they demonstrated the effectiveness of the developed hybrid GA.

Lin and Tsai (2016) developed a deterministic, single objective optimisation algorithm called micro GA to solve the constrained three-dimensional reader network planning problem in the radio frequency identification system. A micro GA with new spatial crossover and correction schemes was developed to solve the problem. Performance of the developed micro GA was compared with Particle Swarm Optimisation algorithm as well as the traditional GA in solving a number of case studies. Experimental data confirmed the effectiveness of the developed micro GA.

Luo et al. (2016) developed a GA to solve an integrated problem of vehicle scheduling and storage location allocation, in container terminal operations. The objective function here was to minimise the berth time of ships. The integrated problem was formulated as a mixed-integer programming model. Experimental data revealed that the developed GA outperforms the commercial software. Again, they did not take uncertainties and multiple objective functions into consideration.

As can be seen from the above literature review, GA has an impressive record and application. A lot of complex real-world problems have been solved by GA. No wonder many versions of GA have been developed and reported in the literature due to the fact that GA is a searching philosophy only; there is no standard GA. Generally speaking, when using GA, users must customise GA components, e.g. chromosome encoding, crossover or mutation to their needs.

2.8 Current Research Gaps

The literature reviewed in Sects. 2.2–2.7 shows that currently there are two major limitations in VCIM production scheduling. First, the existing VCIM production scheduling models are not comprehensive because the collaborative shipment scheduling, multiple objectives, multiple product orders, and uncertainties are not taken into consideration. Second, the existing optimisation methods for solving the VCIM production scheduling problem are not sufficiently robust since not all sub-problems are simultaneously solved. Without simultaneously solving all of the sub-problems, it is impossible to achieve the global optimal solution.

This research aims to overcome these two limitations whose outcome will serve as a foundation towards developing a decision support system capable of helping decision makers to operate VCIM systems more effectively.

2.9 Concluding Remarks

In this chapter, a comprehensive literature review on production scheduling in VCIM systems was presented. The review showed that VCIM is a relatively new global manufacturing system, which has two unique characteristics, i.e. integration and temporary cooperation. Production scheduling is vital to VCIM systems because it affects the quality, cost, and lead time of products. In spite of considerable research in this area, currently there are two major limitations in VCIM production scheduling, i.e. lack of a comprehensive scheduling model and lack of a robust optimisation method. This research aims to overcome these two limitations along with a comprehensive VCIM production scheduling model and a robust Genetic Algorithm.

References

Abd, K., Abhary, K., & Marian, R. (2016). Multi-objective optimisation of dynamic scheduling in robotic flexible assembly cells via fuzzy-based Taguchi approach. *Computers & Industrial Engineering, 99,* 250–259.

Ackooij, W. V., & Malick, J. (2016). Decomposition algorithm for large-scale two-stage unit-commitment. *Annals of Operations Research, 238*(1), 587–613.

Aguilar-Rivera, R., Valenzuela-Rendón, M., & Rodríguez-Ortiz, J. J. (2015). Genetic algorithms and Darwinian approaches in financial applications: A survey. *Expert Systems with Applications, 42*(21), 7684–7697.

Aiello, G., Scalia, L. G., & Enea, M. (2013). A non dominated ranking multi-objective genetic algorithm and electre method for unequal area facility layout problems. *Expert Systems with Applications, 40*(12), 4812–4819.

Akpınar, S., Bayhan, G. M., & Baykasoglu, A. (2013). Hybridizing ant colony optimization via genetic algorithm for mixed-model assembly line balancing problem with sequence dependent setup times between tasks. *Applied Soft Computing, 13*(1), 574–589.

Akram, K., Kamal, K., & Zeb, A. (2016). Fast simulated annealing hybridized with quenching for solving job shop scheduling problem. *Applied Soft Computing, 49,* 510–523.

Asawasakulsorn, A. (2009). Transportation collaboration: Partner selection criteria and IOS design issues for supporting trust. *International Journal of Business and Information, 4*(2), 199–220.

Bal, M., Manesh, H. F., & Hashemipour, M. (2008, April 1). Virtual-reality-based information requirements analysis tool for CIM system implementation: A case study in die-casting industry. *International Journal of Computer Integrated Manufacturing, 21*(3), 231–244.

Balakrishnan, J., Cheng, C. H., Conway, D. G., & Lau, C. M. (2003). A hybrid genetic algorithm for the dynamic plant layout problem. *International Journal of Production Economics, 86*(2), 107–120.

Batur, G. D., Karasan, O. E., & Akturk, M. S. (2012). Multiple part-type scheduling in flexible robotic cells. *International Journal of Production Economics, 135*(2), 726–740.

Boender, C. G. E., & Romeijn, H. E. (1995). Stochastic methods. In R. Horst & P. M. Pardalos (Eds.), *Handbook of global optimization*. Boston: Kluwer Academic Publishers.

Boudissa, E., & Bounekhla, M. (2012). Genetic algorithm with dynamic selection based on quadratic ranking applied to induction machine parameters estimation. *Electric Power Components and Systems, 40*(10), 1089–1104.

Bozdağ, C. E., Kahraman, C., & Ruan, D. (2003). Fuzzy group decision making for selection among computer integrated manufacturing systems. *Computers in Industry, 51*(1), 13–29.

Camarinha-Matos, L. M., & Afsarmanesh, H. (1999). The virtual enterprise concept. In L. M. Camarinha-Matos & H. Afsarmanesh (Eds.), *Infrastructures for virtual enterprises—Networking industrial enterprises* (pp. 3–14). The Netherlands: Kluwer Academic Publishers.

Cardin, O., Mebarki, N., & Pinot, G. (2013). A study of the robustness of the group scheduling method using an emulation of a complex FMS. *International Journal of Production Economics, 146*(1), 199–207.

Castelli, M., & Vanneschi, L. (2014). Genetic algorithm with variable neighborhood search for the optimal allocation of goods in shop shelves. *Operations Research Letters, 42*(5), 355–360.

Castro, H., Putnik, G. D., Shah, V., & Cruz-Cunha, M. M. (2013). A simulation tool and its role in supporting the management of the transformation processes of traditional enterprises into virtual enterprises. *Tékhne, In Press, Corrected Proof.*

Chan, F. T. S., & Zhang, T. (2011). The impact of collaborative transportation management on supply chain performance: A simulation approach. *Expert Systems with Applications, 38*(3), 2319–2329.

Chang, P. C., Hsieh, J. C., & Wang, C. Y. (2007). Adaptive multi-objective genetic algorithms for scheduling of drilling operation in printed circuit board industry. *Applied Soft Computing, 7*(3), 800–806.

Chen, C., Xia, J., Liu, J., & Feng, G. (2006). Nonlinear inversion of potential-field data using a hybrid-encoding genetic algorithm. *Computers & Geosciences, 32*(2), 230–239.

Cheng, B. W., & Chang, C. L. (2007). A study on flowshop scheduling problem combining Taguchi experimental design and genetic algorithm. *Expert Systems with Applications, 32*(2), 415–421.

Coelho, L. D. S., Ayala, H. V. H., & Mariani, V. C. (2014). A self-adaptive chaotic differential evolution algorithm using gamma distribution for unconstrained global optimization. *Applied Mathematics and Computation, 234,* 452–459.

Crispim, J., Rego, N., & Jorge, P. D. S. (2015). Stochastic partner selection for virtual enterprises: A chance-constrained approach. *International Journal of Production Research, 53*(12), 3661–3677.

Crispim, J. A., & Sousa, J. P. D. (2009). Partner selection in virtual enterprises: A multi-criteria decision support approach. *International Journal of Production Research, 47*(17), 4791–4812.

Dai, B., & Chen, H. (2009). Mathematical model and solution approach for collaborative logistics in less than truckload (LTL) transportation. In *International Conference on Computers & Industrial Engineering* (pp. 767–772).

Dao, S.D. & Marian, R. (2011c). Modeling and optimisation of precedence-constrained production sequencing and scheduling using multi-objective genetic algorithm. In *Proceedings of the International Conference of Computational Intelligence and Intelligent Systems*, pp. 1027–1032, 6-8 July, London, U.K.

Dao, S. D., Abhary, K., & Marian, R. (2012). Optimisation of resource scheduling in VCIM systems using genetic algorithm. *International Journal of Advanced Research in Artificial Intelligence, 1*(8), 49–56.

Dao, S. D., Abhary, K., & Marian, R. (2014). Optimisation of partner selection and collaborative transportation scheduling in Virtual Enterprises using GA. *Expert Systems with Applications, 41*(15), 6701–6717.

Dao, S. D., Abhary, K., & Marian, R. (2015). An adaptive restarting genetic algorithm for global optimization. In *Proceedings of the World Congress on Engineering and Computer Science* (pp. 455–459).

Dao, S.D., Abhary, K., Marian, R. (2016a). A stochastic production scheduling model for VCIM systems. *Intelligent Industrial Systems*, 2(1), 85-101.

Dao, S. D., Abhary, K., & Marian, R. (2016b). An improved structure of genetic algorithms for global optimisation. *Progress in Artificial Intelligence, 5*(3), 155–163.

Dao, S. D., Abhary, K., & Marian, R. (2017). A bibliometric analysis of Genetic Algorithms throughout the history. *Computers & Industrial Engineering, 110,* 395–403.

Dao, S. D., & Marian, R. (2011a). Optimisation of precedence-constrained production sequencing and scheduling using genetic algorithms. In *International MultiConference of Engineers and Computer Scientists.*

Dao, S. D., & Marian, R. (2011b). Modeling and optimisation of precedence-constrained production sequencing and scheduling for multiple production lines using genetic algorithm. *Computer Technology and Application., 2*(6), 487–499.

Dao, S. D., & Marian, R. (2013). Genetic algorithms for integrated optimisation of precedence-constrained production sequencing and scheduling. In S.-I. Ao & L. Gelman (Eds.), *Electrical engineering and intelligent systems* (pp. 65–80). New York: Springer.

Deep, K., Singh, K. P., Kansal, M. L., & Mohan, C. (2009). A real coded genetic algorithm for solving integer and mixed integer optimization problems. *Applied Mathematics and Computation, 212*(2), 505–518.

Deep, K., & Thakur, M. (2007). A new mutation operator for real coded genetic algorithms. *Applied Mathematics and Computation, 193*(1), 211–230.

Ergun, Ö., Kuyzu, G., & Savelsbergh, M. (2007). Shipper collaboration. *Computers & Operations Research, 34*(6), 1551–1560.

Esen, İ., & Koç, M. A. (2015). Optimization of a passive vibration absorber for a barrel using the genetic algorithm. *Expert Systems with Applications, 42*(2), 894–905.

Faghihi, V., Reinschmidt, K. F., & Kang, J. H. (2014). Construction scheduling using genetic algorithm based on building information model. *Expert Systems with Applications, 41*(16), 7565–7578.

Fahimnia, B., Luong, L., & Marian, R. (2008). Optimization/simulation modeling of the integrated production-distribution plan: An innovative survey. *WSEAS TRANSACTIONS on Business and Economics, 3*(5), 52–65.

Fang, Y., Chu, F., Mammar, S., & Shi, Q. (2015). A new cut-and-solve and cutting plane combined approach for the capacitated lane reservation problem. *Computers & Industrial Engineering, 80,* 212–221.

Floudas, C. A. (2000). *Deterministic global optimization: Theory, methods and applications.* Boston, MA: Springer.

Gao, K. Z., Suganthan, P. N., Pan, Q. K., Chua, T. J., Chong, C. S., & Cai, T. X. (2016). An improved artificial bee colony algorithm for flexible job-shop scheduling problem with fuzzy processing time. *Expert Systems with Applications, 65,* 52–67.

Gen, M., & Cheng, R. (1997). *Genetic algorithms and engineering design.* New York: Wiley.

Goldberg, D. E. (1989). *Genetic algorithms in search, optimization, and machine learning.* Boston: Addison-Wesley Publishing Company, Inc.

Hanagandi, V., & Nikolaou, M. (1998). A hybrid approach to global optimization using a clustering algorithm in a genetic search framework. *Computers & Chemical Engineering, 22* (12), 1913–1925.

Harrington, J. (1973). *Computer integrated manufacturing.* New York: Industrial Press.

He, N., Zhang, D. Z., & Li, Q. (2014). Agent-based hierarchical production planning and scheduling in make-to-order manufacturing system. *International Journal of Production Economics, 149,* 117–130.

He, Y., & Hui, C. W. (2010). A binary coding genetic algorithm for multi-purpose process scheduling: A case study. *Chemical Engineering Science, 65*(16), 4816–4828.

Hfaiedh, W., Sadfi, C., Kacem, I., & Hadj-Alouane, A. (2015). A branch-and-bound method for the single-machine scheduling problem under a non-availability constraint for maximum delivery time minimization. *Applied Mathematics and Computation, 252,* 496–502.

Holland, J. H. (1975). *Adaptation in natural and artificial systems.* Ann Arbor: University of Michigan Press.

Huang, B., Gao, C., & Chen, L. (2011). Partner selection in a virtual enterprise under uncertain information about candidates. *Expert Systems with Applications, 38*(9), 11305–11310.

Huang, B., Gou, H., Liu, W., Li, Y., & Xie, M. (2002). A framework for virtual enterprise control with the holonic manufacturing paradigm. *Computers in Industry, 49*(3), 299–310.

Huang, M., & Fan, C. (2007). Research on the partner selection of virtual enterprise based on self-adaptive genetic algorithm. In *Third International Conference on Intelligent Information Hiding and Multimedia Signal Processing* (pp. 330–333).

Hyun, C. J., Kim, Y., & Kim, Y. K. (1998). A genetic algorithm for multiple objective sequencing problems in mixed model assembly lines. *Computers & Operations Research, 25*(7–8), 675–690.

Ip, W. H., Huang, M., Yung, K. L., & Wang, D. (2003). Genetic algorithm solution for a risk-based partner selection problem in a virtual enterprise. *Computers & Operations Research, 30*(2), 213–231.

Kahraman, C., Beskese, A., & Ruan, D. (2004). Measuring flexibility of computer integrated manufacturing systems using fuzzy cash flow analysis. *Information Sciences, 168*(1–4), 77–94.

Kundakcı, N., & Kulak, O. (2016). Hybrid genetic algorithms for minimizing makespan in dynamic job shop scheduling problem. *Computers & Industrial Engineering, 96,* 31–51.

Lan, F., Bayraksan, G., & Lansey, K. (2016). Reformulation linearization technique based branch-and-reduce approach applied to regional water supply system planning. *Engineering Optimization, 48*(3), 454–475.

Lee, K., Kim, B. S., & Joo, C. M. (2012). Genetic algorithms for door-assigning and sequencing of trucks at distribution centers for the improvement of operational performance. *Expert Systems with Applications, 39*(17), 12975–12983.

Li, J., & Chan, F. T. S. (2012). The impact of collaborative transportation management on demand disruption of manufacturing supply chains. *International Journal of Production Research, 50* (19), 5635–5650.

Li, J., Huang, Y., & Niu, X. (2016). A branch population genetic algorithm for dual-resource constrained job shop scheduling problem. *Computers & Industrial Engineering, 102,* 113–131.

Liberti, L., & Kucherenko, S. (2005). Comparison of deterministic and stochastic approaches to global optimization. *International Transactions in Operational Research, 12*(3), 263–285.

Lin, G. C. I. (1997). The latest research trends in CIM. In *The Fourth International Conference on Computer Integrated Manufacturing* (pp. 26–33).

Lin, S. Y., & Tsai, H. F. (2016). Micro genetic algorithm with spatial crossover and correction schemes for constrained three-dimensional reader network planning. *Expert Systems with Applications, 44,* 344–353.

Lin, W. C., Yin, Y., Cheng, S. R., Cheng, T. C. E., Wu, C. H., & Wu, C. C. (2016). Particle swarm optimization and opposite-based particle swarm optimization for two-agent multi-facility customer order scheduling with ready times. *Applied Soft Computing*. https://doi.org/10.1016/j.asoc.2016.09.038.

Luo, J., Wu, Y., & Mendes, A. B. (2016). Modelling of integrated vehicle scheduling and container storage problems in unloading process at an automated container terminal. *Computers & Industrial Engineering, 94,* 32–44.

Mahmoodabadi, M. J., Safaie, A. A., Bagheri, A., & Nariman-zadeh, N. (2013). A novel combination of Particle Swarm Optimization and Genetic Algorithm for Pareto optimal design of a five-degree of freedom vehicle vibration model. *Applied Soft Computing, 13*(5), 2577–2591.

Maity, S., Roy, A., & Maiti, M. (2015). A modified genetic algorithm for solving uncertain constrained solid travelling salesman problems. *Computers & Industrial Engineering, 83,* 273–296.

Majumder, A., & Laha, D. (2016). A new cuckoo search algorithm for 2-machine robotic cell scheduling problem with sequence-dependent setup times. *Swarm and Evolutionary Computation, 28,* 131–143.

Marian, R.M., Luong, L. & Dao, S.D. (2012). Hybrid genetic algorithm optimisation of distribution networks—A comparative study, in SI Ao, O Castillo & X Huang (eds), *Intelligent Control and Innovative Computing*, Springer US, Boston, MA, pp. 109-122.

Michalewicz, Z. (1996). *Genetic algorithms + Data structures = Evolution programs*. Berlin, Heidelberg: Springer.

Miller, F. P., Vandome, A. F., & McBrewster, J. (2010). *Computer-integrated manufacturing*. Mauritius: VDM Publishing House.

Moles, C. G., Mendes, P., & Banga, J. R. (2003). Parameter estimation in biochemical pathways: A comparison of global optimization methods. *Genome Research, 13*(11), 2467–2474.

Mun, J., Shin, M., Lee, K., & Jung, M. (2009). Manufacturing enterprise collaboration based on a goal-oriented fuzzy trust evaluation model in a virtual enterprise. *Computers & Industrial Engineering, 56*(3), 888–901.

Nagalingam, S. V. (1999). *An innovative decision support system for CIM justification and optimisation*. Ph.D. thesis, School of Engineering, University of South Australia.

Nagalingam, S. V., & Lin, G. C. I. (1999). Latest developments in CIM. *Robotics and Computer-Integrated Manufacturing, 15*(6), 423–430.

Nagalingam, S. V., & Lin, G. C. I. (2000). A distributed group decision support systems for virtual computer integrated manufacturing. In *The 6th International Conference on Automation Technology* (pp. 377–382).

Nagalingam, S. V., Lin, G. C. I., & Wang, D. (2007). Resource scheduling for a virtual CIM system. In L. Wang & W. Shen (Eds.), *Process planning and scheduling for distributed manufacturing* (pp. 269–294). London: Springer.

Nagalingam, S. V., Lin, G. C. I., Zhou, J., & Wang, D. (2003). Virtual computer integrated manufacturing and its future applications. In *Proceedings of the 17th International Conference on Production Research.*

Ng, C. K., & Li, D. (2014). Test problem generator for unconstrained global optimization. *Computers & Operations Research, 51,* 338–349.
Niu, S. H., Ong, S. K., & Nee, A. Y. C. (2012). An enhanced ant colony optimiser for multi-attribute partner selection in virtual enterprises. *International Journal of Production Research, 50*(8), 2286–2303.
Ozener, O. O. (2008). *Collaboration in transportation.* Ph.D. thesis, Georgia Institute of Technology, Atlanta.
Qing-dao-er-ji, R., & Wang, Y. (2012). A new hybrid genetic algorithm for job shop scheduling problem. *Computers & Operations Research, 39*(10), 2291–2299.
Qu, H., Xing, K., & Alexander, T. (2013). An improved genetic algorithm with co-evolutionary strategy for global path planning of multiple mobile robots. *Neurocomputing, 120,* 509–517.
Salehi, M., & Moghaddam, T. R. (2009). Application of genetic algorithm to computer-aided process planning in preliminary and detailed planning. *Engineering Applications of Artificial Intelligence, 22*(8), 1179–1187.
Sawik, T. (2007). A lexicographic approach to bi-objective scheduling of single-period orders in make-to-order manufacturing. *European Journal of Operational Research, 180*(3), 1060–1075.
Shahlaei, M., Sobhani, A. M., Saghaie, L., & Fassihi, A. (2012). Application of an expert system based on Genetic Algorithm–Adaptive Neuro-Fuzzy Inference System (GA–ANFIS) in QSAR of cathepsin K inhibitors. *Expert Systems with Applications, 39*(6), 6182–6191.
Shahvari, O., & Logendran, R. (2017). An enhanced tabu search algorithm to minimize a bi-criteria objective in batching and scheduling problems on unrelated-parallel machines with desired lower bounds on batch sizes. *Computers & Operations Research, 77,* 154–176.
Shokouhifar, M., & Jalali, A. (2015a). A new evolutionary based application specific routing protocol for clustered wireless sensor networks. *International Journal of Electronics and Communications (AEÜ), 69,* 432–441.
Shokouhifar, M., & Jalali, A. (2015b). An evolutionary-based methodology for symbolic simplification of analog circuits using genetic algorithm and simulated annealing. *Expert Systems with Applications, 42*(3), 1189–1201.
Simona, D., & Raluca, P. (2011). Intelligent modeling method based on genetic algorithm for partner selection in virtual organizations. *Business and Economic Horizons, 5*(2), 23–34.
Stern, H., Chassidim, Y., & Zofi, M. (2006). Multiagent visual area coverage using a new genetic algorithm selection scheme. *European Journal of Operational Research, 175*(3), 1890–1907.
Su, W., Huang, S. X., Fan, Y. S., & Mak, K. L. (2015). Integrated partner selection and production–distribution planning for manufacturing chains. *Computers & Industrial Engineering, 84,* 32–42.
Suresh, S., Huang, H., & Kim, H. J. (2014). Hybrid real-coded genetic algorithm for data partitioning in multi-round load distribution and scheduling in heterogeneous systems. *Applied Soft Computing, 24,* 500–510.
Tang, P. H., & Tseng, M. H. (2013). Adaptive directed mutation for real-coded genetic algorithms. *Applied Soft Computing, 13*(1), 600–614.
Tao, F., Zhang, L., Zhang, Z. H., & Nee, A. Y. C. (2010). A quantum multi-agent evolutionary algorithm for selection of partners in a virtual enterprise. *CIRP Annals - Manufacturing Technology, 59*(1), 485–488.
Wan, Y., Wang, M., Ye, Z., & Lai, X. (2016). A feature selection method based on modified binary coded ant colony optimization algorithm. *Applied Soft Computing, 49,* 248–258.
Wang, D. (2007). *The development of an agent-based architecture for virtual CIM.* Ph.D. thesis, University of South Australia, Adelaide.
Wang, D., Nagalingam, S. V., & Lin, G. C. I. (2003a). Development of a virtual CIM system using agent-based approach. In *Proceedings of the 7th International Conference on Mechatronics Technology* (pp. 445–450).
Wang, D., Nagalingam, S. V., & Lin, G. C. I. (2003b). Development of an optimised virtual CIM system. In *The 17th International Conference on Production Research* (pp. 1–5).

Wang, D., Nagalingam, S. V., & Lin, G. C. I. (2004). Development of a parallel processing multi-agent architecture for a virtual CIM system. *International Journal of Production Research, 42*(17), 3765–3785.

Wang, D., Nagalingam, S. V., & Lin, G. C. I. (2007). Development of an agent-based virtual CIM architecture for small to medium manufacturers. *Robotics and Computer-Integrated Manufacturing, 23*(1), 1–16.

Wang, L., & Zheng, D. Z. (2002). A modified genetic algorithm for job shop scheduling. *The International Journal of Advanced Manufacturing Technology, 20*(1), 72–76.

Wang, N. F., Zhang, X. M., & Yang, Y. W. (2013a). A hybrid genetic algorithm for constrained multi-objective optimization under uncertainty and target matching problems. *Applied Soft Computing, 13*(8), 3636–3645.

Wang, Y., Huang, J., Dong, W. S., Yan, J. C., Tian, C. H., Li, M., et al. (2013b). Two-stage based ensemble optimization framework for large-scale global optimization. *European Journal of Operational Research, 228*(2), 308–320.

Wang, Y., Yin, H., & Wang, J. (2009). Genetic algorithm with new encoding scheme for job shop scheduling. *The International Journal of Advanced Manufacturing Technology, 44*(9–10), 977–984.

Wozabal, D. (2012). Value-at-risk optimization using the difference of convex algorithm. *OR Spectrum, 34*(4), 861–883.

Wu, N., & Su, P. (2005). Selection of partners in virtual enterprise paradigm. *Robotics and Computer-Integrated Manufacturing, 21*(2), 119–131.

Wu, X., Chu, C. H., Wang, Y., & Yan, W. (2007). A genetic algorithm for cellular manufacturing design and layout. *European Journal of Operational Research, 181*(1), 156–167.

Xia, H., Li, X., & Gao, L. (2016). A hybrid genetic algorithm with variable neighborhood search for dynamic integrated process planning and scheduling. *Computers & Industrial Engineering, 102,* 99–112.

Yang, C. O., Guan, T. Y., & Lin, J. S. (2000). Developing a computer shop floor control model for a CIM system—using object modeling technique. *Computers in Industry, 41*(3), 213–238.

Ye, F. (2010). An extended TOPSIS method with interval-valued intuitionistic fuzzy numbers for virtual enterprise partner selection. *Expert Systems with Applications, 37*(10), 7050–7055.

Ye, F., & Li, Y. N. (2009). Group multi-attribute decision model to partner selection in the formation of virtual enterprise under incomplete information. *Expert Systems with Applications, 36*(5), 9350–9357.

Yolmeh, A., & Salehi, N. (2015). An outer approximation method for an integration of supply chain network designing and assembly line balancing under uncertainty. *Computers & Industrial Engineering, 83,* 297–306.

Yu, C., Xu, X., & Lu, Y. (2015). Computer-integrated manufacturing, cyber-physical systems and cloud manufacturing—Concepts and relationships. *Manufacturing Letters, 6,* 5–9.

Yun, Y., Chung, H., & Moon, C. (2013). Hybrid genetic algorithm approach for precedence-constrained sequencing problem. *Computers & Industrial Engineering, 65*(1), 137–147.

Yurdakul, M. (2004). Selection of computer-integrated manufacturing technologies using a combined analytic hierarchy process and goal programming model. *Robotics and Computer-Integrated Manufacturing, 20*(4), 329–340.

Yusuf, İ. T. (2012). An experimental design approach using TOPSIS method for the selection of computer-integrated manufacturing technologies. *Robotics and Computer-Integrated Manufacturing, 28*(2), 245–256.

Zeng, Z. B., Li, Y., & Zhu, W. (2006). Partner selection with a due date constraint in virtual enterprises. *Applied Mathematics and Computation, 175*(2), 1353–1365.

Zhang, Y., Tao, F., Laili, Y., Hou, B., Lv, L., & Zhang, L. (2012). Green partner selection in virtual enterprise based on Pareto genetic algorithms. *The International Journal of Advanced Manufacturing Technology*, 1–17.

Zhang, Y., Tao, F., Laili, Y., Hou, B., Lv, L., & Zhang, L. (2013). Green partner selection in virtual enterprise based on Pareto genetic algorithms. *The International Journal of Advanced Manufacturing Technology, 67*(9), 2109–2125.

Zhao, J., & Wang, L. (2011). Center based genetic algorithm and its application to the stiffness equivalence of the aircraft wing. *Expert Systems with Applications, 38*(5), 6254–6261.

Zhao, Q., Zhang, X., & Xiao, R. (2008). Particle swarm optimization algorithm for partner selection in virtual enterprise. *Progress in Natural Science, 18*(11), 1445–1452.

Zhong, T. X., & Chen, J. C. (2002). A hybrid-coded genetic algorithm based optimisation of non-productive paths in CNC machining. *The International Journal of Advanced Manufacturing Technology, 20*(3), 163–168.

Zhong, Y., Jian, L., & Zijun, W. (2009). An integrated optimisation algorithm of GA and ACA-based approaches for modeling virtual enterprise partner selection. *The DATA BASE for Advances in Information Systems, 40*(2), 37–56.

Zhou, N., Xing, K., & Nagalingam, S. V. (2010a). An agent-based cross-enterprise resource planning for small and medium enterprises. *IAENG International Journal of Computer Science, 37*(3), 1–7.

Zhou, N., Xing, K., Nagalingam, S. V., & Lin, G. C. I. (2010b). Development of an agent based VCIM resource scheduling process for small and medium enterprises. In *Proceedings of the International MultiConference of Engineers and Computer Scientists* (pp. 39–44).

Zhou, W., Zheng, J., Yan, J., & Wang, J. (2011). A novel hybrid algorithm for assembly sequence planning combining bacterial chemotaxis with genetic algorithm. *The International Journal of Advanced Manufacturing Technology, 52*(5–8), 715–724.

Chapter 3
An Innovative Production Scheduling Model for VCIM Systems

3.1 Introduction

As indicated in Chap. 2, the VCIM production scheduling, an important issue in operation of any VCIM system, is a dynamic constrained optimisation problem. Currently, there are two main limitations in the VCIM production scheduling: lack of a comprehensive scheduling model and a robust optimisation solution method. To overcome the limitation of existing models, an innovative production scheduling model for VCIM systems is proposed in this chapter which is a stochastic model, not only capable of handling multiple customer orders but also multiple objective functions. Additionally, in the proposed model, two major sub-problems of the VCIM production scheduling, namely partner selection and collaborative shipment scheduling, are fully integrated together to facilitate the global optimisation.

3.2 Proposed Production Scheduling Model for VCIM Systems

3.2.1 Problem Statement

Generally, after receiving product orders, the VCIM system decomposes the orders into a number of components which can be independently produced; and then the VCIM system does the production scheduling by (1) selecting some component suppliers to produce the required components, (2) selecting some assembly agents to assemble the required products and (3) scheduling the shipments to transport the components and products to the required destinations; so that a temporary VCIM production system can be formed to fulfil the customer orders. To take advantage of collaborative shipment and to facilitate the global optimisation, an innovative production scheduling model for VCIM systems is developed herein as follows.

S. D. Dao, *Modelling and Intelligent Optimisation of Production Scheduling in VCIM Systems*, Springer Theses, https://doi.org/10.1007/978-3-319-72113-2_3

Consider:

- A VCIM system has a number of assembly agents as well as component suppliers, which are both globally and locally distributed.
- The VCIM system is capable of providing a number of products for customers worldwide and each product is made by assembling a number of components.
- Each component supplier can provide certain components for certain products.
- Each assembly agent can do final assembly for certain products.
- In the current planning horizon, several customer orders with the same and/or different products are being requested.

Determine:

- Allocation of the required components of the requested products to the component suppliers.
- Allocation of the requested products to the assembly agents.
- The related component and product shipping schedules.

So that:

A temporary VCIM production system can be formed to simultaneously fulfil all of the customer orders with (1) minimum total cost, (2) minimum average completion time and (3) maximum average reliability, while all of the given constraints are satisfied.

Conditions:

- Cost and release time of each product component provided by different component suppliers are different from each another but known in advance.
- All product components can be independently produced.
- Assembly cost and time of each product assembled in different assembly agents are also different but given in advance.
- Shipping cost and time between any two locations are different from each other but known in advance.
- All component release times, product assembly times and shipping times are stochastic parameters with known distribution functions, mean values as well as standard deviations.
- Reliabilities of the components/products, which are defined as probabilities of producing the components/products with qualities satisfying the pre-set standards, made by different component suppliers/assembly agents are different but known in advance.
- Assembly of any single product of any type is done by one assembly agent only.
- One assembly agent is capable of assembling a certain number of products.
- All component suppliers and assembly agents have enough resources such as labour, material, etc. to perform their designed functions 24 h a day, 7 days a week.

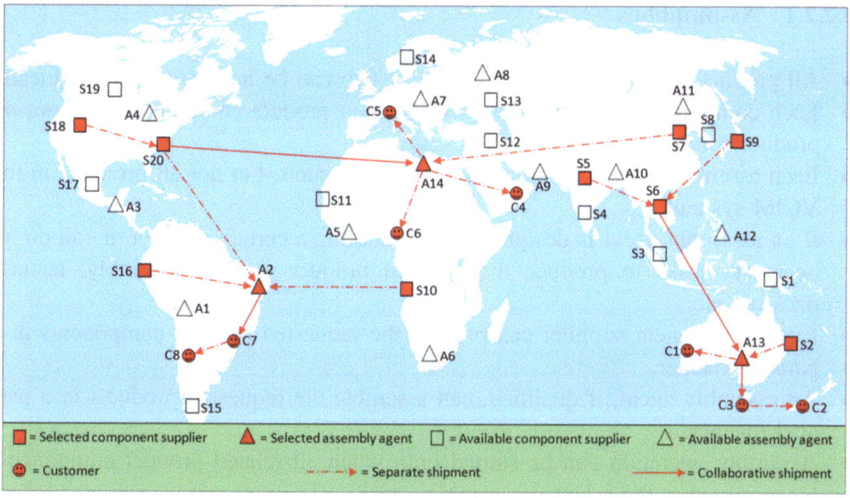

Fig. 3.1 The proposed VCIM production scheduling model (Dao et al. 2016)

The proposed VCIM production scheduling model is illustrated in Fig. 3.1 with a typical production scheduling solution to fulfil 8 customer orders (C1, C2 … C8). The VCIM system illustrated in Fig. 3.1 consists of 14 assembly agents (A1, A2 … A14) and 20 component suppliers (S1, S2 … S20), distributed worldwide. In the illustrated solution, component suppliers S2, S5, S6, S9 and assembly agent A13 were selected to serve customers C1–C3; component suppliers S7, S18, S20 and assembly agent A14 were selected to fulfil the customer orders C4–C6; and component suppliers S10, S16, S20 and assembly agent A2 were selected to serve customers C7–C8. Two kinds of shipments, separate shipment and collaborative shipment, were used in the solution. It is noted that the separate shipment herein refers to a shipment carrying the component(s) produced by only one component supplier or a shipment carrying finished product(s) to only one customer. In contrast, the collaborative shipment means a shipment carrying the components made by more than one component supplier or a shipment carrying finished products to more than one customer. Examples of separate and collaborative shipments are shown in Fig. 3.1. It is noted that this problem statement has been published in a conference paper (Dao et al. 2016).

3.2.2 Model Formulation

In order to describe the proposed VCIM production scheduling model in detail, the following mathematical model is developed.

3.2.2.1 Assumptions

- All product components in the VCIM system can be independently produced.
- Each component supplier can provide certain product components, but not all product components in the VCIM system.
- Each assembly agent can assemble certain products, but not all products in the VCIM system.
- If an assembly agent is designed for assembling a certain product, it can do all necessary tasks to produce the finished product such as assembly, testing, packing, etc.
- Every component supplier can process the requested product components in a parallel manner.
- An assembly agent, if qualified, can assemble the requested products in a parallel manner.
- Assembly operation can be started only when all related product components have arrived at the selected assembly agent.
- One item of a product is assembled in one assembly agent only.
- Each customer orders only one item of one product at a time.
- All component release times, product assembly times as well as shipping times are stochastic parameters following normal distributions.
- Reliabilities of the components/products, which are defined as probabilities of producing the components/products with qualities satisfying the pre-set standards, made by different component suppliers/assembly agents are different but known in advance.
- There is only one shipment available from one place to another in one planning horizon.
- Shipping cost consists of fixed shipping cost and variable shipping cost.
- The fixed shipping cost is proportional to the shipping distance that is the length of the line segment between two locations. It should be noted that the fixed shipping cost is a constant because the distance between any two locations is a constant.
- The variable shipping cost is proportional to the items' actual weight or cubic weight, whichever is greater. Cubic weight is a measure of size/volume of items transported, and the formula used to calculate the cubic weight will be presented in Sect. 3.4.
- All component suppliers as well as assembly agents have enough resources such as material, labour, etc. to perform their designed functions 24 h a day, 7 days a week.

3.2.2.2 Indices

i = Component index
j = Product index
s = Component supplier index

a = Assembly agent index
c = Customer index.

3.2.2.3 Deterministic Input Parameters

A = Number of assembly agents in the VCIM system
A_j^a = Assembly cost of product j assembled in assembly agent a ($)
AA_j^a = Binary parameter representing the availability of assembly operation, which is 1 if assembly agent a is capable of assembling product j; 0 otherwise
AW_j = Actual weight of product j (kg)
AW_{ij} = Actual weight of component i of product j (kg)
C = Number of customers in the current planning horizon
C_{ij}^s = Cost of component i of product j provided by component supplier s ($)
CA_{ij}^s = Binary parameter representing the availability of product component, which is 1 if component supplier s is capable of providing component i of product j; 0 otherwise
CW_{ij} = Cubic weight of component i of product j (kg)
CW_j = Cubic weight of product j (kg)
$D_{s_1s_2}$ = Distance between component suppliers s_1 and s_2 (km)
D_{sa} = Distance between component supplier s and assembly agent a (km)
D_{ac} = Distance between assembly agent a and customer c (km)
$D_{c_1c_2}$ = Distance between customers c_1 and c_2 (km)
$DPMO_{ij}^s$ = Reliability of component i of product j, provided by component supplier s, in term of defects per million opportunities (DPMO). DPMO is a very popular quality index in Total Quality Management and Six Sigma.
$DPMO_j^a$ = Reliability, in term of DPMO, of assembly operation in assembly agent a for product j
F = Fixed shipping cost coefficient
N_j = Number of components of product j
P = Number of products under consideration
S = Number of component suppliers in the VCIM system
V = Variable shipping cost coefficient.

3.2.2.4 Stochastic Input Parameters

$AS_{j_2}^a \sim N\left(\mu_{j_2}^a, \sigma_{j_2}^a\right)$ = Assembly time of product j_2 at assembly agent a. This is a stochastic parameter following normal distribution with mean value $\mu_{j_2}^a$ and standard deviation $\sigma_{j_2}^a$ (day)

$RT_{i_1 j_1}^{s_1} \sim N\left(\mu_{i_1 j_1}^{s_1}, \sigma_{i_1 j_1}^{s_1}\right)$ = Release time of component i_1 of product j_1, provided by component supplier s_1. This is a stochastic parameter following normal distribution with mean value $\mu_{i_1 j_1}^{s_1}$ and standard deviation $\sigma_{i_1 j_1}^{s_1}$ (day)

$ST^{s_1 s_2} \sim N(\mu^{s_1 s_2}, \sigma^{s_1 s_2})$ = Shipping time from component supplier s_1 to component supplier s_2. This is a stochastic parameter following normal distribution with mean value $\mu^{s_1 s_2}$ and standard deviation $\sigma^{s_1 s_2}$ (day)

$ST^{s_1 a} \sim N(\mu^{s_1 a}, \sigma^{s_1 a})$ = Shipping time from component supplier s_1 to assembly agent a. This is a stochastic parameter following normal distribution with mean value $\mu^{s_1 a}$ and standard deviation $\sigma^{s_1 a}$ (day)

$ST^{ac} \sim N(\mu^{ac}, \sigma^{ac})$ = Shipping time from assembly agent a to customer c. This is a stochastic parameter following normal distribution with mean value μ^{ac} and standard deviation σ^{ac} (day)

$ST^{c_1 c_2} \sim N(\mu^{c_1 c_2}, \sigma^{c_1 c_2})$ = Shipping time from customer c_1 to customer c_2. This is a stochastic parameter following normal distribution with mean value $\mu^{c_1 c_2}$ and standard deviation $\sigma^{c_1 c_2}$ (day).

3.2.2.5 Intermediate Parameters

$AT_{i_1 j_1}^{s_0 s_1}$ = Arrival time at component supplier s_1 of component i_1 of product j_1, shipped from component supplier s_0 (day)

$DT_{i_2 j_2}^{s_1 s_2}$ = Dispatch time of component i_2 of product j_2, shipped from component supplier s_1 to component supplier s_2 (day)

$DT_{i_2 j_2}^{s_1 a}$ = Dispatch time of component i_2 of product j_2, shipped from component supplier s_1 to assembly agent a (day)

$AT_{i_2 j_2}^{s_1 a}$ = Arrival time at assembly agent a of component i_2 of product j_2, shipped from component supplier s_1 (day)

$DT_{j_2}^{ac}$ = Dispatch time of product j_2, shipped from assembly agent a to customer c (day)

$AT_{j_2}^{ac}$ = Arrival time of product j_2, shipped from assembly agent a to customer c (day).

3.2.2.6 Special Mathematical Functions

$$Fn\{X\} = \begin{cases} 1 & \text{If X} > 0 \\ 0 & \text{Otherwise} \end{cases}$$

$$Max\{X;Y\} = \begin{cases} X & \text{If X} \geq \text{Y} \\ Y & \text{Otherwise} \end{cases}$$

$$Sn\{X\} = \begin{cases} 1 & \text{If X} = 0 \\ 0 & \text{Otherwise} \end{cases}$$

3.2.2.7 Decision Variables

$$X_{ij}^{s} = \begin{cases} 1 & \text{If component supplier } s \text{ is selected to provide component } i \text{ of product } j \\ 0 & \text{Otherwise} \end{cases}$$

$$Y_{j}^{a} = \begin{cases} 1 & \text{If assembly agent } a \text{ is selected to assemble product } j \\ 0 & \text{Otherwise} \end{cases}$$

$$Z_{ij}^{s_1 s_2} = \begin{cases} 1 & \text{If component } i \text{ of product } j \text{ is shipped from component supplier } s_1 \text{ to component supplier } s_2 \\ 0 & \text{Otherwise} \end{cases}$$

$$Z_{ij}^{sa} = \begin{cases} 1 & \text{If component } i \text{ of product } j \text{ is shipped from component supplier } s \text{ to assembly agent } a \\ 0 & \text{Otherwise} \end{cases}$$

$$Z_{j}^{ac} = \begin{cases} 1 & \text{If product } j \text{ is shipped from assembly agent } a \text{ to customer } c \\ 0 & \text{Otherwise} \end{cases}$$

$$Z_{j}^{c_1 c_2} = \begin{cases} 1 & \text{If product } j \text{ is shipped from customer } c_1 \text{ to customer } c_2 \\ 0 & \text{Otherwise.} \end{cases}$$

3.2.2.8 Multiple Objective Functions

The proposed VCIM production scheduling model considers three objective functions, namely minimum total cost, minimum average completion time and maximum average reliability. The three objective functions are calculated as follows.

Objective function 1: Total cost (*TC*) of all product orders under consideration, which is the sum of total component cost (*CC*), total assembly cost (*AC*) and total shipping cost (*SC*):

$$TC = CC + AC + SC \tag{3.1}$$

where *CC, AC* and *SC* are calculated as follows:

- Total component cost (*CC*):

$$CC = \sum_{s=1}^{S} \sum_{j=1}^{P} \sum_{i=1}^{N_j} C_{ij}^{s} X_{ij}^{s} \tag{3.2}$$

- Total assembly cost (*AC*) of all products under consideration:

$$AC = \sum_{a=1}^{A} \sum_{j=1}^{P} A_j^a Y_j^a \tag{3.3}$$

- Total shipping cost (*SC*) required to fulfil all product orders under consideration, which is the sum of the following four cost components: (1) total cost of shipping the product components from one selected component supplier to another (SC_1), (2) total cost of shipping the product components from the selected component suppliers to the selected assembly agents (SC_2), (3) total cost of shipping the finished products from the selected assembly agents to the customers who have direct shipment contacts with the selected assembly agents (SC_3) and (4) total cost of shipping the finished products from one customer to another (SC_4).

$$SC = SC_1 + SC_2 + SC_3 + SC_4 \tag{3.4}$$

where SC_1, SC_2, SC_3 and SC_4 are calculated as follows:

- Total cost of shipping the product components from one selected component supplier to another (SC_1), which is the sum of the fixed cost (first term) and variable cost (second term), as follows:

$$\begin{aligned} SC_1 = & \sum_{s_2=1}^{S} \sum_{s_1=1}^{S} \left[D_{s_1 s_2}.F.Fn\left\{ \sum_{j=1}^{P} \sum_{i=1}^{N_j} Z_{ij}^{s_1 s_2} \right\} \right] \\ & + \sum_{s_2=1}^{S} \sum_{s_1=1}^{S} \left[V.\left(Max\left\{ \sum_{j=1}^{P} \sum_{i=1}^{N_j} AW_{ij}.Z_{ij}^{s_1 s_2}; \sum_{j=1}^{P} \sum_{i=1}^{N_j} CW_{ij}.Z_{ij}^{s_1 s_2} \right\} \right) \right] \end{aligned} \tag{3.5}$$

- Total cost of shipping the product components from the selected component suppliers to the selected assembly agents (SC_2), again, is the sum of the fixed cost (first term) and variable cost (second term), as follows:

$$\begin{aligned} SC_2 = & \sum_{a=1}^{A} \sum_{s=1}^{S} \left[D_{sa}.F.Fn\left\{ \sum_{j=1}^{P} \sum_{i=1}^{N_j} Z_{ij}^{sa} \right\} \right] \\ & + \sum_{a=1}^{A} \sum_{s=1}^{S} \left[V.\left(Max\left\{ \sum_{j=1}^{P} \sum_{i=1}^{N_j} AW_{ij}.Z_{ij}^{sa}; \sum_{j=1}^{P} \sum_{i=1}^{N_j} CW_{ij}.Z_{ij}^{sa} \right\} \right) \right] \end{aligned} \tag{3.6}$$

- Total cost of shipping the finished products from the selected assembly agents to the customers who have direct shipment contacts with the selected assembly agents (SC_3), again, is the sum of the fixed cost (first term) and variable cost (second term), as follows:

$$SC_3 = \sum_{c=1}^{C}\sum_{a=1}^{A}\left[D_{ac}.F.Fn\left\{\sum_{j=1}^{P} Z_j^{ac}\right\}\right] + \sum_{c=1}^{C}\sum_{a=1}^{A}\left[V.\left(Max\left\{\sum_{j=1}^{P} AW_j.Z_j^{ac};\sum_{j=1}^{P} CW_j.Z_j^{ac}\right\}\right)\right] \tag{3.7}$$

- Total cost of shipping the finished products from one customer to another (SC_4), once again, is the sum of the fixed cost (first term) and variable cost (second term), as follows:

$$SC_4 = \sum_{c_2=1}^{C}\sum_{c_1=1}^{C}\left[D_{c_1c_2}.F.Fn\left\{\sum_{j=1}^{P} Z_j^{c_1c_2}\right\}\right] + \sum_{c_2=1}^{C}\sum_{c_1=1}^{C}\left[V.\left(Max\left\{\sum_{j=1}^{P} AW_j.Z_j^{c_1c_2};\sum_{j=1}^{P} CW_j.Z_j^{c_1c_2}\right\}\right)\right] \tag{3.8}$$

Objective function 2: Average completion time of all product orders under consideration (CO). In order to determine the completion times of the product orders, a number of intermediate time components need to be sequentially calculated, because of sequential nature of the time slots, as follows.

- Dispatch time of component i_2 of product j_2, shipped from component supplier s_1 to component supplier s_2 ($DT_{i_2j_2}^{s_1s_2}$) and arrival time of component i_2 of product j_2, shipped from component supplier s_1 to component supplier s_2 ($AT_{i_2j_2}^{s_1s_2}$). It is noted that the two time components are interlinked to each other; and at one shipping node, i.e. one selected component supplier, arrival times of components involved must be calculated before their dispatch times; but at two different shipping nodes, the calculating sequence of the arrival/dispatch times cannot be fixed. General equations to calculate $DT_{i_2j_2}^{s_1s_2}$ and $AT_{i_2j_2}^{s_1s_2}$ are as follows:

$$\begin{cases} DT_{i_2j_2}^{s_1s_2} = \max\limits_{\forall s_0,i_1,j_1}\left\{\left[RT_{i_1j_1}^{s_1} \sim N\left(\mu_{i_1j_1}^{s_1},\sigma_{i_1j_1}^{s_1}\right)\right].X_{i_1j_1}^{s_1}.Z_{i_1j_1}^{s_1s_2};\ AT_{i_1j_1}^{s_0s_1}.Z_{i_1j_1}^{s_1s_2}\right\} \\ Z_{i_2j_2}^{s_1s_2} = 1 \end{cases}$$
$$\forall s_1, s_2, i_2, j_2 \tag{3.9}$$

$$\begin{cases} AT_{i_2j_2}^{s_1s_2} = DT_{i_2j_2}^{s_1s_2} + ST^{s_1s_2} \sim N(\mu^{s_1s_2},\sigma^{s_1s_2}) \\ Z_{i_2j_2}^{s_1s_2} = 1 \end{cases} \quad \forall s_1, s_2, i_2, j_2 \tag{3.10}$$

It should be noted that each general equation, either Eq. 3.9 or 3.10, contains a series of equations, as many as the number of times that the condition $Z_{i_2j_2}^{s_1s_2} = 1$ is satisfied; which means that the dispatch/arrival times $DT_{i_2j_2}^{s_1s_2}/AT_{i_2j_2}^{s_1s_2}$ will be calculated if and only if the shipping decision variable is equal to 1 ($Z_{i_2j_2}^{s_1s_2} = 1$). In addition, after all $DT_{i_2j_2}^{s_1s_2}$ and $AT_{i_2j_2}^{s_1s_2}$ are calculated, arrival times of all product components transported to the component suppliers, who have direct shipment contacts with the selected assembly agents, are determined. The next step is to calculate dispatch times of the product components transported from the component suppliers who have direct shipment contacts with the selected assembly agents to the selected assembly agents. Then the product component arrival times at the selected assembly agents are calculated.

- Dispatch time of component i_2 of product j_2, shipped from component supplier s_1 to assembly agent a ($DT_{i_2j_2}^{s_1a}$):

$$\begin{cases} DT_{i_2j_2}^{s_1a} = \max\limits_{\forall s_0, i_1, j_1} \left\{ \left[RT_{i_1j_1}^{s_1} \sim N\left(\mu_{i_1j_1}^{s_1}, \sigma_{i_1j_1}^{s_1}\right) \right] .X_{i_1j_1}^{s_1}.Z_{i_1j_1}^{s_1a};\ AT_{i_1j_1}^{s_0s_1}.Z_{i_1j_1}^{s_1a} \right\} \\ Z_{i_2j_2}^{s_1a} = 1 \end{cases} \quad \forall s_1, a, i_2, j_2 \tag{3.11}$$

- Arrival time of component i_2 of product j_2, shipped from component supplier s_1 to assembly agent a ($AT_{i_2j_2}^{s_1a}$):

$$\begin{cases} AT_{i_2j_2}^{s_1a} = DT_{i_2j_2}^{s_1a} + ST^{s_1a} \sim N(\mu^{s_1a}, \sigma^{s_1a}) \\ Z_{i_2j_2}^{s_1a} = 1 \end{cases} \quad \forall s_1, a, i_2, j_2 \tag{3.12}$$

- Dispatch time of product j_2, shipped from assembly agent a to customer c ($DT_{j_2}^{ac}$):

$$\begin{cases} DT_{j_2}^{ac} = \max\limits_{\forall j_2} \left\{ \max\limits_{\forall s_1, i_2} \left\{ AT_{i_2j_2}^{s_1a} \right\} + AS_{j_2}^{a} \sim N\left(\mu_{j_2}^{a}, \sigma_{j_2}^{a}\right) \right\} \\ Y_{j_2}^{a} = 1 \\ Z_{j_2}^{ac} = 1 \end{cases} \quad \forall j_2, a, c \tag{3.13}$$

- Arrival time of product j_2, shipped from assembly agent a to customer c ($AT_{j_2}^{ac}$):

$$\begin{cases} AT_{j_2}^{ac} = DT_{j_2}^{ac} + ST^{ac} \sim N(\mu^{ac}, \sigma^{ac}) \\ Z_{j_2}^{ac} = 1 \end{cases} \quad \forall j_2, a, c. \tag{3.14}$$

- Completion time of customer order with product j_2 (CT_{j_2}):

$$\begin{cases} CT_{j_2} = AT_{j_2}^{ac} + \sum_{c_2=1}^{C} \sum_{c_1=1}^{C} \left[[ST^{c_1c_2} \sim N(\mu^{c_1c_2}, \sigma^{c_1c_2})].Z_{j_2}^{c_1c_2} \right] \\ Z_{j_2}^{ac} = 1 \end{cases} \quad \forall j_2, a, c \tag{3.15}$$

- Average completion time of all customer orders under consideration (CO):

$$CO = \frac{\sum_{j_2=1}^{P} CT_{j_2}}{P} \tag{3.16}$$

Objective function 3: Average reliability of all products under consideration (AR):

$$AR = \frac{\sum_{j=1}^{P} (RP_j)}{P} \tag{3.17}$$

where RP_j is the reliability of product j, calculated as follows:

$$RP_j = \prod_{\forall i,s} \left(1 - \frac{DPMO_{ij}^{s}.X_{ij}^{s}}{10^6}\right) . \prod_{\forall a} \left(1 - \frac{DPMO_{j}^{a}.Y_{j}^{a}}{10^6}\right), \quad \forall j \tag{3.18}$$

3.2.2.9 Constraints

- Component availability:

$$CA_{ij}^{s} \geq X_{ij}^{s}, \quad \forall i, j, s \tag{3.19}$$

- Assembly operation availability:

$$AA_{j}^{a} \geq Y_{j}^{a}, \quad \forall j, a \tag{3.20}$$

- Component supplier selection:

$$\sum_{s=1}^{S} X_{ij}^{s} = 1, \quad \forall i,j \tag{3.21}$$

- Assembly agent selection:

$$\sum_{a=1}^{A} Y_{j}^{a} = 1, \quad \forall j \tag{3.22}$$

- Shipping flow conservation of product component(s) at component supplier s_2:

$$\sum_{s_1=1}^{S}\sum_{j=1}^{P}\sum_{i=1}^{N_j} Z_{ij}^{s_1 s_2} + \sum_{j=1}^{P}\sum_{i=1}^{N_j} X_{ij}^{s_2} = \sum_{s_3=1}^{S}\sum_{j=1}^{P}\sum_{i=1}^{N_j} Z_{ij}^{s_2 s_3} + \sum_{a=1}^{A}\sum_{j=1}^{P}\sum_{i=1}^{N_j} Z_{ij}^{s_2 a}, \quad \forall s_2 \tag{3.23}$$

- Product components belonging to one product order once transported together in one shipment, should not be split to be transported in different shipments. That is because: (1) these product components will eventually arrive at the same assembly agent, (2) splitting the components will make the computation very complicated and (3) the related shipping cost is more likely to be higher, since more fixed shipping cost may be involved. However, product components belonging to different product orders, once transported together in one shipment, can be split, since the components may have different destinations. The following constraint ensures that product components belonging to one product order, once transported together in one shipment, cannot be split:

$$\sum_{s_2=1}^{S}\left(Fn\left\{ \sum_{i=1}^{N_j} Z_{ij}^{s_1 s_2} \right\} \right) \le 1, \quad \forall j, s_1 \tag{3.24}$$

- All components of product j must arrive at the selected assembly agent a:

$$\sum_{s=1}^{S}\sum_{i=1}^{N_j} Z_{ij}^{sa} = N_j . Y_{j}^{a}, \quad \forall j, a \tag{3.25}$$

- Only component suppliers, who are selected to provide product components, are involved in the component shipping schedule:

$$Fn\left\{\sum_{j=1}^{P}\sum_{i=1}^{N_j} X_{ij}^{s_2}\right\} \geq Fn\left\{\sum_{s_1=1}^{S}\sum_{j=1}^{P}\sum_{i=1}^{N_j} Z_{ij}^{s_1 s_2}\right\}, \quad \forall s_2 \tag{3.26}$$

- Total number of products shipped from assembly agents to customers:

$$\sum_{c=1}^{C}\sum_{a=1}^{A}\sum_{j=1}^{P} Z_j^{ac} = P \tag{3.27}$$

- Shipping flow conservation of products at customer c_2:

$$\sum_{c_1=1}^{C}\sum_{j=1}^{P} Z_j^{c_1 c_2} = \sum_{c_3=1}^{C}\sum_{j=1}^{P} Z_j^{c_2 c_3} + 1, \quad \forall c_2 \tag{3.28}$$

- Product shipping route without branches (applied to the shipping routes between the selected assembly agents and the customers who have direct shipment contacts with the assembly agents):

$$\sum_{a=1}^{A}\left(Fn\left\{\sum_{j=1}^{P} Z_j^{ac}\right\}\right) \leq 1, \quad \forall c \tag{3.29}$$

- Product shipping route without branches (applied to the shipping routes between customers):

$$\sum_{c_2=1}^{C}\left(Fn\left\{\sum_{j=1}^{P} Z_j^{c_1 c_2}\right\}\right) \leq 1, \quad \forall c_1 \tag{3.30}$$

$$\sum_{c_1=1}^{C}\left(Fn\left\{\sum_{j=1}^{P} Z_j^{c_1 c_2}\right\}\right) \leq 1, \quad \forall c_2 \tag{3.31}$$

- Shipping right products to right customers:

$$\sum_{c=1}^{C}\sum_{a=1}^{A}\sum_{j=1}^{P}\left[Z_j^{ac}.Sn\{j-c\}\right]+\sum_{c_2=1}^{C}\sum_{c_1=1}^{C}\sum_{j=1}^{P}\left[Z_j^{c_1c_2}.Sn\{j-c_2\}\right]=P \tag{3.32}$$

- Valid values of the indices and parameters:

$$i, i_1, i_2 = 1, 2, 3, \ldots, N_j \tag{3.33}$$

$$j, j_1, i_2 = 1, 2, 3, \ldots, P \tag{3.34}$$

$$s, s_0, s_1, s_2, s_3 = 1, 2, 3, \ldots, S \tag{3.35}$$

$$a = 1, 2, 3, \ldots, A \tag{3.36}$$

$$c, c_1, c_2, c_3 = 1, 2, 3, \ldots, C \tag{3.37}$$

$$N_j, P, S, A, C = positiveintegers \tag{3.38}$$

$$CA_{ij}^s, AA_j^a = binarynumbers \tag{3.39}$$

$$C_{ij}^s, A_j^a, F, D_{s_1s_2}, AW_{ij}, CW_{ij}, V, D_{sa}, D_{ac}, AW_j, CW_j, D_{c_1c_2}, DPMO_{ij}^s > 0 \tag{3.40}$$

$$DPMO_j^a, RT_{i_1j_1}^{s_1} \sim N\left(\mu_{i_1j_1}^{s_1}, \sigma_{i_1j_1}^{s_1}\right), ST^{s_1s_2} \sim N(\mu^{s_1s_2}, \sigma^{s_1s_2}) > 0 \tag{3.41}$$

$$ST^{s_1a} \sim N(\mu^{s_1a}, \sigma^{s_1a}), AS_{j_2}^a \sim N\left(\mu_{j_2}^a, \sigma_{j_2}^a\right) > 0 \tag{3.42}$$

$$ST^{ac} \sim N(\mu^{ac}, \sigma^{ac}), ST^{c_1c_2} \sim N(\mu^{c_1c_2}, \sigma^{c_1c_2}) > 0 \tag{3.43}$$

It should be noted that this is the most comprehensive VCIM production scheduling model ever developed to date. Computational complexity and some aspects of the proposed model will be discussed in the next sections.

3.3 Computational Complexity

As can be seen from Sect. 3.2.2, the scheduling problem in the proposed model is a dynamic multi-objective optimisation problem under uncertainties. The scheduling problem is dynamic because number of partners involved, i.e. the selected component suppliers and assembly agents, may vary from solution to solution. In addition, the scheduling problem consists of two sub-problems, namely partner selection and collaborative shipment scheduling, which need to be simultaneously solved to facilitate the global optimisation. It is noted that the partner selection sub-problem herein refers to as selecting component suppliers and assembly agents.

Combinatorial optimisation is a class of computational problems dealing with the questions of selecting the optimum from a finite set of possibilities (Lovász 2005). Obviously, the problem in the proposed VCIM production scheduling model is a combinatorial optimisation problem. Finding the global optimal solutions to combinatorial optimisation problems is not trivial. Although it is easy to deal with small-size problems, it is very hard, if not impossible, when coping with large-size or real-world problems (Dowlatshahi et al. 2014).

From computational complexity theory point of view, the scheduling problem in the proposed model is a NP-hard problem, because of the following reasons. First, this scheduling problem would become a Job Shop Scheduling problem (Gao et al. 2015; Kuhpfahl and Bierwirth 2016) if the collaborative shipment is not considered; a typical solution is illustrated in Fig. 3.2. Second, the scheduling problem would become an Open Vehicle Routing problem (López-Sánchez et al. 2014; Salari et al. 2010) if the initial problem would be simplified when the partners, i.e. the selected component suppliers and assembly agents, are fixed and no shipping route with branches is allowed; a typical solution is illustrated in Fig. 3.3. In other words, two well-known combinatorial optimisation problems, Job Shop Scheduling and Open Vehicle Routing, are the special cases of the scheduling problem in the proposed VCIM production scheduling model. Finally, both Job Shop Scheduling and Open Vehicle Routing problems are NP-hard problems (Gao et al. 2015; Kurdi 2015; López-Sánchez et al. 2014; Salari et al. 2010). Therefore, it can be concluded that the scheduling problem in the proposed VCIM production scheduling model belongs to the complexity class NP-hard, and it is much more difficult to solve, compared to very well-known Job Shop Scheduling and Open Vehicle Routing problems.

It should be noted that NP is a technical term in computational complexity theory in computer science and mathematics, which stands for Non-deterministic

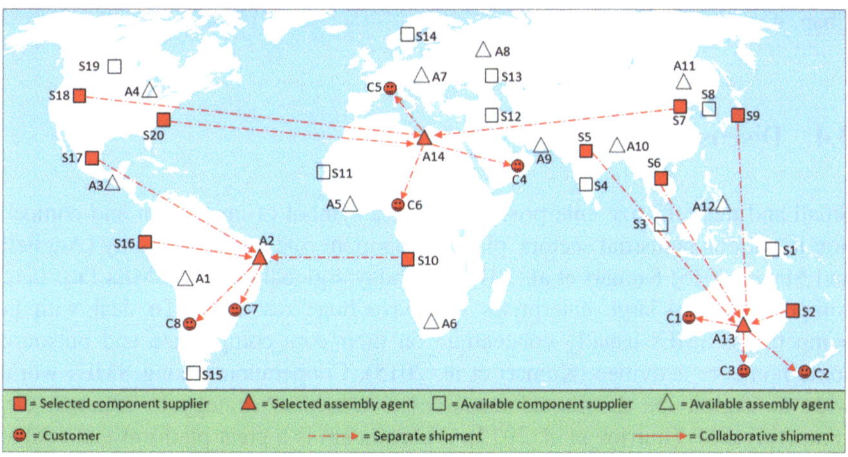

Fig. 3.2 A typical VCIM production schedule without collaborative shipment

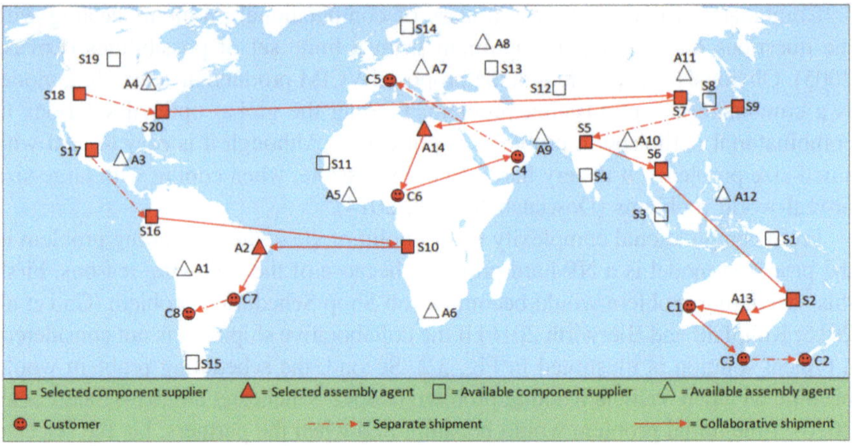

Fig. 3.3 A typical VCIM production schedule with no-branch shipping route

Polynomial-time. NP problems are the set of decision problems that can be solved by non-deterministic polynomial-time bounded Turing machines (Cadoli et al. 2000). In addition, NP-hard is the class of decision problems which are as hard as any NP problem (Shapiro and Delgado-Eckert 2012). NP-hard problems are algorithmically solvable but computationally intractable (Shapiro and Delgado-Eckert 2012). There is no exact method that can find the global optimal solutions to NP-hard problems in polynomial time, and fast approximate heuristics and meta-heuristics are the popular approaches to search for high-quality/practical solutions (He et al. 2012). In this thesis, Genetic Algorithms (GA) based approach is used to solve the scheduling problem in the proposed VCIM production scheduling model. The reason of choosing GA among other meta-heuristics was already justified in Chap. 2 and details of the developed GA will be presented in Chap. 4.

3.4 Discussions

Small and medium size enterprises (SMEs), a symbol of innovation and competition in various industrial sectors, play an important role in any economy (Angilella and Mazzù 2015; Kumari et al. 2015). In today's global market, SMEs face fierce competition from large enterprises that have huge resources. To deal with the competition, SMEs usually concentrate on their core competence and outsource some non-core activities (Kumari et al. 2015). Cooperation among SMEs worldwide to exploit the distributed resources is a critical strategy to enhance their competitiveness (Iturrioz et al. 2015). Alibaba.com is a great platform that enables

SMEs to do business efficiently anywhere in the world (Yazdanifard and Li 2014). The philosophy that makes Alibaba Company so successful is as follows:

> Small and medium-size enterprises are like grains of sand on a beach. The Internet can glue them together. It can make them into an invincible force that is able to go up against the big stones. The virtue of the Internet is that it uses small to compete with big, it uses fast to compete with slow.
>
> Jack Ma, the founder of e-commerce giant Alibaba (Liu and Avery 2009, p. 50).

Alibaba.com is the e-commerce giant, running B2B (business to business), B2C (business to customer) and C2C (customer to customer) e-marketplace that connects small and medium-size buyers and suppliers all over the world (Qing and Xue 2009). Jack Ma, a former English teacher, started Alibaba Company in 1999 with just $60,000. But now, Alibaba's profits surpass those of Ebay and Amazon combined; only three technology companies in the world, i.e. Microsoft, Google and Apple, are bigger (Waite 2014).

With Alibaba.com, barriers all around the world dissolve; anyone with Internet access has the opportunity to import, export, build commercial alliances, and trade with anyone else, anywhere in the world. In addition, customers can have a high degree of participation in the design of the products they purchase (Schepp and Schepp 2010). There is no doubt that the Alibaba.com platform can support VCIM implementation in general and the proposed VCIM production scheduling model in particular, in which component suppliers and assembly agents could be SMEs distributed worldwide.

VCIM is a global manufacturing system, in which component suppliers, assembly agents and customers are locally as well as globally distributed. Therefore, the component/product shipping distances in the system are very large and complex; obtaining the exact distances is not practical. That is why the shipping distances used in this study are assumed to be the direct distances between places, calculated based on the corresponding Latitudes and Longitudes.

Temporary cooperation is one of the unique characteristics of VCIM systems. Component suppliers and assembly agents in a VCIM system are SMEs, which run their own businesses and cooperate with each other when activated by the VCIM system. One component supplier can provide a number of product components at the same time because it may have them in store and may not have to produce them right after the request. In the worst case scenario in which a component supplier has to set up production lines to produce some components, the component release times will be accordingly adjusted and updated in the VCIM database. The same tactic is applied to assembly times of products at assembly agents; i.e. in the worst case scenario if an assembly agent cannot assemble the products in parallel, the assembly times will be adjusted. Thereby, the assumptions regarding parallel processing capabilities of component suppliers and assembly agents are acceptable. Thanks to the parallel processing capability assumptions, computational burden required to solve the scheduling problem can be reduced.

Assuming stochastic manufacturing, assembly and shipping times following normal distribution is quite popular in the operations research literature (Axsäter

2003; Chang et al. 2009; Ehmke et al. 2015; Hoque 2013; Horng et al. 2012; Hung and Hsiao 2013; Lemos and Ronconi 2015; Liao and Chang 2011; Tiacci 2015). However, different systems may have different characteristics; to determine the real distribution types of stochastic parameters of a system, analysing statistical data collected from the real system must be done. In VCIM systems, collecting such kind of statistical data becomes very easy after the systems have operated for some time, because everything is recorded in the VCIM database. Therefore, the distribution type of each stochastic parameter of the system will be updated accordingly. At this stage, for the sake of simplicity, release times of product components, assembly times of products, shipping times of components/products in the proposed VCIM production scheduling model are stochastic parameters assumed to follow normal distribution.

As can be seen from the proposed VCIM production scheduling model, product component shipping routes can have a number of branches, for example, the shipping route connecting component suppliers S_5, S_6, S_9 and assembly agent A_{13} as illustrated in Fig. 3.1. However, product shipping routes are assumed to have no branches, for instance, the shipping routes: $A_2 \rightarrow C_7 \rightarrow C_8$ and $A_{13} \rightarrow C_3 \rightarrow C_2$ as shown in Fig. 3.1. That is because of the following reasons. First, one product component made by one component supplier can stay or wait at another component supplier according to the collaborative shipping plan approved by the VCIM system. For example, product components made by S_5 and S_9 can stay or wait at S_6 for the combined shipment to A_{13} as illustrated in Fig. 3.1. Second, unlike the components, it is not legal if the finished product belonging to one customer stays or waits at another customer for another shipment, i.e. the product shipping route must be continuous, like the routes $A_2 \rightarrow C_7 \rightarrow C_8$ and $A_{13} \rightarrow C_3 \rightarrow C_2$ in Fig. 3.1.

All components and products in a VCIM system are standardised, so that they can be made by different partners, i.e. different component suppliers and assembly agents. Trying to produce components/products with quality exceeding the pre-set standard is not necessary. Therefore, the term "reliability", instead of the term "quality", is used in the third objective function in the proposed model. The reliability of a product in this study is the probability of producing the product with quality satisfying the pre-set standard, which is calculated through DPMO indices of the corresponding components and assembly operation as shown in Eq. 3.18. It should be noted that DPMO, standing for defects per million opportunities, is a popular quality index used in many books and research papers associated with Quality Management and Six Sigma (Antony et al. 2006; Yang and El-Haik 2003).

In the proposed VCIM production scheduling model, shipping cost between two locations consists of two cost components, namely fixed cost and variable cost. The fixed cost is proportional to the shipping distance. The variable cost is proportional to actual weight or cubic weight, whichever is greater. The cubic weight is a measure of size/volume of item(s), which is used by Australia_Post (2015) and calculated as shown in Fig. 3.4.

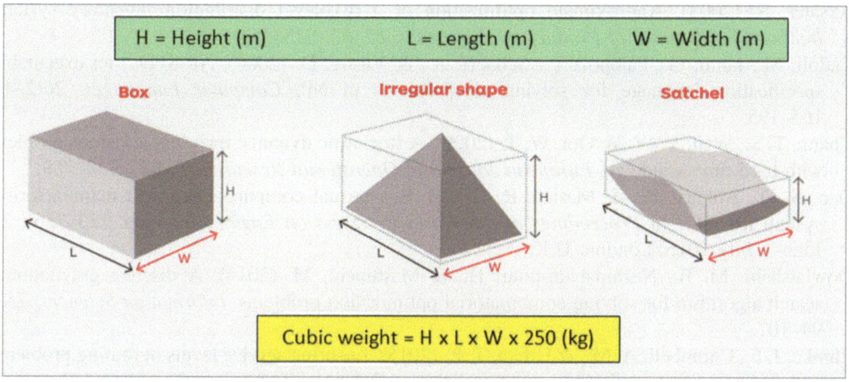

Fig. 3.4 Cubic weight calculation (Australia_Post 2015)

3.5 Concluding Remarks

In this chapter, an innovative production scheduling model for VCIM systems has been proposed. The developed model, the most comprehensive VCIM production scheduling model to date, is not only capable of handling multiple product orders simultaneously but also multiple objective functions, under uncertainties of component release times, assembly times as well as shipping times. In addition, besides selecting appropriate partners, i.e. component suppliers and assembly agents, to fulfil product orders, the proposed model can support collaborative shipment scheduling.

The proposed model has been explicitly expressed in terms of mathematical equations. The scheduling problem in the proposed VCIM production scheduling model is a dynamic multi-objective optimisation problem with uncertainties. From computational complexity theory point of view, this scheduling problem is a NP-hard problem. Moreover, a number of assumptions in the proposed model have been discussed and clarified. Finally, it has been concluded that Alibaba.com platform, run by e-commerce giant Alibaba, can support implementing the proposed VCIM production scheduling model.

References

Angilella, S., & Mazzù, S. (2015). The financing of innovative SMEs: a multicriteria credit rating model. *European Journal of Operational Research, 244*(2), 540–554.

Antony, J., Kumar, A., & Banuelas, R. (2006). *World class applications of six sigma*. Hoboken: Taylor and Francis.

Australia_Post (2015). *Size & weight guidelines*, Australia Post. Accessed October 06, 2015. (http://auspost.com.au/parcels-mail/size-and-weight-guidelines.html?ilink=mm-size-and-weight-guidelines-1).

Axsäter, S. (2003). Approximate optimization of a two-level distribution inventory system. *International Journal of Production Economics, 81–82,* 545–553.

Cadoli, M., Ianni, G., Palopoli, L., Schaerf, A., & Vasile, D. (2000). NP-SPEC: an executable specification language for solving all problems in NP. *Computer Languages, 26*(2–4), 165–195.

Chang, T. S., Wan, Y. W., & Ooi, W. T. (2009). A stochastic dynamic traveling salesman problem with hard time windows. *European Journal of Operational Research, 198*(3), 748–759.

Dao, S. D., Abhary, K., & Marian, R. (2016). Is a virtual computer-integrated manufacturing system feasible? In *Proceedings of the World Congress on Engineering*, (pp. 733–736), 29 June–1 July 2016, London, U.K.

Dowlatshahi, M. B., Nezamabadi-pour, H., & Mashinchi, M. (2014). A discrete gravitational search algorithm for solving combinatorial optimization problems. *Information Sciences, 258,* 94–107.

Ehmke, J. F., Campbell, A. M., & Urban, T. L. (2015). Ensuring service levels in routing problems with time windows and stochastic travel times. *European Journal of Operational Research, 240*(2), 539–550.

Gao, L., Li, X., Wen, X., Lu, C., & Wen, F. (2015). A hybrid algorithm based on a new neighborhood structure evaluation method for job shop scheduling problem. *Computers and Industrial Engineering, 88,* 417–429.

He, K., Huang, W., & Jin, Y. (2012). An efficient deterministic heuristic for two-dimensional rectangular packing. *Computers and Operations Research, 39*(7), 1355–1363.

Hoque, M. A. (2013). A vendor–buyer integrated production–inventory model with normal distribution of lead time. *International Journal of Production Economics, 144*(2), 409–417.

Horng, S. C., Lin, S. S., & Yang, F. Y. (2012). Evolutionary algorithm for stochastic job shop scheduling with random processing time. *Expert Systems with Applications, 39*(3), 3603–3610.

Hung, Y. F., & Hsiao, J. Y. (2013). Inventory rationing decision models during replenishment lead time. *International Journal of Production Economics, 144*(1), 290–300.

Iturrioz, C., Aragón, C., & Narvaiza, L. (2015). How to foster shared innovation within SMEs' networks: social capital and the role of intermediaries. *European Management Journal, 33*(2), 104–115.

Kuhpfahl, J., & Bierwirth, C. (2016). A study on local search neighborhoods for the job shop scheduling problem with total weighted tardiness objective. *Computers and Operations Research, 66,* 44–57.

Kumari, S., Singh, A., Mishra, N., & Garza-Reyes, J. A. (2015). A multi-agent architecture for outsourcing SMEs manufacturing supply chain. *Robotics and Computer-Integrated Manufacturing, 36,* 36–44.

Kurdi, M. (2015). A new hybrid island model genetic algorithm for job shop scheduling problem. *Computers and Industrial Engineering, 88,* 273–283.

Lemos, R. F., & Ronconi, D. P. (2015). Heuristics for the stochastic single-machine problem with E/T costs. *International Journal of Production Economics, 168,* 131–142.

Liao, H. C., & Chang, H. H. (2011). The optimal approach for parameter settings based on adjustable contracting capacity for the hospital supply chain logistics system. *Expert Systems with Applications, 38*(5), 4790–4797.

Liu, S., & Avery, M. (2009). *Alibaba: the inside story behind Jack Ma and the creation of the world's biggest online marketplace*. New York: HarperCollins e-books.

López-Sánchez, A. D., Hernández-Díaz, A. G., Vigo, D., Caballero, R., & Molina, J. (2014). A multi-start algorithm for a balanced real-world open vehicle routing problem. *European Journal of Operational Research, 238*(1), 104–113.

Lovász, L. (2005). Review of the book by Alexander Schrijver: combinatorial optimization: polyhedra and efficiency. *Operations Research Letters, 33*(4), 437–440.

Qing, H. H., & Xue, Z. S. (2009). A model for value-added e-marketplace provisioning: case study from Alibaba.com. In C. Godart et al. (Eds.), *Software Services for e-business and e-society* (Vol. 305, pp. 65–72). Berlin Heidelberg: Springer.

Salari, M., Toth, P., & Tramontani, A. (2010). An ILP improvement procedure for the open vehicle routing problem. *Computers and Operations Research, 37*(12), 2106–2120.
Schepp, B., & Schepp, D. (2010). *The official Alibaba.com success guide insider tips and strategies for sourcing products from the world's largest B2B marketplace*. Hoboken: Wiley.
Shapiro, M., & Delgado-Eckert, E. (2012). Finding the probability of infection in an SIR network is NP-Hard. *Mathematical Biosciences, 240*(2), 77–84.
Tiacci, L. (2015). Coupling a genetic algorithm approach and a discrete event simulator to design mixed-model un-paced assembly lines with parallel workstations and stochastic task times. *International Journal of Production Economics, 159,* 319–333.
Waite, S. (2014). Record-breaking Alibaba allocates to happy few. *FinanceAsia (Deal of the month: October)* 8–9. http://ezlibproxy.unisa.edu.au/login?, http://search.ebscohost.com/login.aspx?direct=true&db=bth&AN=99088865&site=ehost-live.
Yang, K., & El-Haik, B. (2003). *Design for Six Sigma: a roadmap for product development.* New York: McGraw-Hill.
Yazdanifard, R., & Li, M. T. H. (2014). The review of Alibaba's online business marketing strategies which navigate them to present success. *Global Journal of Management and Business Research, 7*(9), 33–39.

Chapter 4
An Innovative GA for VCIM Production Scheduling Optimisation

4.1 Introduction

As indicated in Chap. 2, the VCIM production scheduling, an important issue to operation of any VCIM system, is a dynamic constrained optimisation problem. Currently, there are two main limitations in the VCIM production scheduling, namely lack of a comprehensive scheduling model and lack of a robust optimisation method. To overcome the first limitation, a new production scheduling model for VCIM systems was developed in Chap. 3. This is the most comprehensive VCIM production scheduling model to date, which is not only capable of handling multiple product orders simultaneously but also multiple objective functions, under uncertainties. In addition, the developed model is capable of supporting collaborative shipment scheduling. It was shown in Chap. 3 that the scheduling problem in the developed VCIM production scheduling model is a dynamic multi-objective optimisation problem with uncertainties; and from the view point of computational complexity theory, this is an NP-hard problem. To overcome the second limitation as well as to solve the scheduling problem in the VCIM production scheduling model developed in Chap. 3, an innovative Genetic Algorithm (GA) is proposed in this chapter. It should be noted that the reason of choosing GA among other meta-heuristics has already been justified in Chap. 2.

With innovative algorithm structure, chromosome encoding, crossover and mutation, the proposed GA is capable of searching for optimal/suboptimal solutions to the large-scale complex scheduling problem in the developed VCIM production scheduling model. There are seven parts in the proposed GA, namely *chromosome encoding*, *crossover*, *mutation*, *evaluation*, *selection*, *algorithm structure* and *parameter tuning*.

S. D. Dao, *Modelling and Intelligent Optimisation of Production Scheduling in VCIM Systems*, Springer Theses, https://doi.org/10.1007/978-3-319-72113-2_4

4.2 Chromosome Encoding

To encode a solution to the scheduling problem in the developed VCIM production scheduling model, each chromosome in the proposed GA has four parts representing four sub-solutions. They are *component supplier selection*, *assembly agent selection*, *component shipment scheduling* and *product shipment scheduling*. Without loss of generality, it is assumed herein that a VCIM has 12 component suppliers and 8 assembly agents; also 4 customer orders are currently under consideration.

The first two parts of a chromosome dealing with partner selection sub-problems are illustrated in Table 4.1. The first part, highlighted in green, represents the component supplier selection. The second part, highlighted in yellow, represents the assembly agent selection. In this example, product orders O_1, O_2, O_3, and O_4 are decomposed into 8, 7, 10 and 9 independent product components (C_i), respectively. The positive integer values in cells highlighted in green represent the corresponding component suppliers selected to provide the corresponding product components (C_i) of the corresponding product orders (O_i). For example, component supplier 4 is selected to provide component C_1 of product order O_1. In the second part of a chromosome, the last column in Table 4.1, the positive integer values represent the corresponding assembly agents chosen for final assembly operations of the corresponding product orders. For instance, assembly agent 2 is selected for final assembly operation of product order O_1. It should be noted that different product orders can be decomposed into different numbers of product components, e.g. orders O_1 and O_2 are decomposed into 8 and 7 product components, respectively, as shown in Table 4.1. In addition, one component supplier can be selected to provide more than one product component and one assembly agent can be selected for final assembly operation of more than one product order, for example, component suppliers 2, 4, 5, 6, 12, and assembly agent 2, as illustrated in Table 4.1. As a result, the number of partners involved in different chromosome may not be the same.

The third part of a chromosome is about shipment scheduling to transport the product components from the selected component suppliers to the selected assembly agents to produce the requested products, as illustrated in Table 4.2. There are four matrices corresponding to the component shipping schedules of four product orders (O_1–O_4). Each component shipping schedule involves all component suppliers (S_1–S_{12}) and one corresponding assembly agent (A). It should be

Table 4.1 First two parts of a chromosome—partner selection

	C_1	C_2	C_3	C_4	C_5	C_6	C_7	C_8	C_9	C_{10}	A
O_1	4	12	3	2	11	6	10	2			2
O_2	6	5	9	3	2	4	4				6
O_3	3	12	5	10	7	5	3	5	3	7	6
O_4	4	1	7	11	7	10	10	5	4		2

Table 4.2 Third part of a chromosome—component shipment scheduling

	S_1	S_2	S_3	S_4	S_5	S_6	S_7	S_8	S_9	S_{10}	S_{11}	S_{12}	A
S_1													
S_2													1
S_3		1											
S_4			1										
S_5													
S_6											1		
S_7													
S_8													
S_9													
S_{10}						1							
S_{11}													1
S_{12}			1										

O_1

	S_1	S_2	S_3	S_4	S_5	S_6	S_7	S_8	S_9	S_{10}	S_{11}	S_{12}	A
S_1													
S_2													1
S_3									1				
S_4			1										
S_5													1
S_6													1
S_7													
S_8													
S_9													1
S_{10}													
S_{11}													
S_{12}													

O_2

	S_1	S_2	S_3	S_4	S_5	S_6	S_7	S_8	S_9	S_{10}	S_{11}	S_{12}	A
S_1													
S_2													
S_3							1						
S_4													
S_5							1						
S_6													
S_7													1
S_8													
S_9													
S_{10}							1						
S_{11}													
S_{12}							1						

O_3

	S_1	S_2	S_3	S_4	S_5	S_6	S_7	S_8	S_9	S_{10}	S_{11}	S_{12}	A
S_1													1
S_2													
S_3													
S_4	1												
S_5										1			
S_6													
S_7													1
S_8													
S_9													
S_{10}											1		
S_{11}													1
S_{12}													

O_4

noted that the assembly agent (A) involved in each component shipping schedule is the one that was selected in the second part, for example, assembly agents 2 and 6 are involved in the component shipping schedules O_1 and O_2, respectively.

A shipping decision to transport the product component(s) from one selected component supplier to another is represented by a binary number: "1" meaning that there is a shipment between the two corresponding component suppliers in the direction determined by "left-right-up" rule and "0" meaning otherwise. For example, in the component shipping schedule O_1, as shown in Table 4.2, there is a shipment from component supplier S_3 to component supplier S_2. For the sake of the table simplicity, all of the binary values of "0" for the shipping decisions are replaced by blank cells in this thesis. In addition, a binary value in column A represents a shipping decision to transport the product component(s) from the corresponding component supplier to the corresponding assembly agent. For instance, in the component shipping schedule O_1 in Table 4.2, there is a shipment from component supplier S_2 to assembly agent 2.

It should be noted that column A in each component shipping schedule (O_i) may have more than one non-zero element, schedule O_2 for example, because one assembly agent may receive the product components from a number of component suppliers. In addition, in each row associated with the component supplier that was selected in the first part of a chromosome, there is one element with value of "1"; and in other rows, every element is "0", which means that the corresponding component suppliers are not involved in the component shipping schedules. In other words, only component suppliers that were selected in the first part of a chromosome are involved in the component shipping schedules. Moreover, the third part of a chromosome is partially connected to the first and second ones. That is because component suppliers and assembly agents, involved in the third part, are determined in the first two parts of a chromosome. The component shipping schedules illustrated in Table 4.2 are visualised in Fig. 4.1.

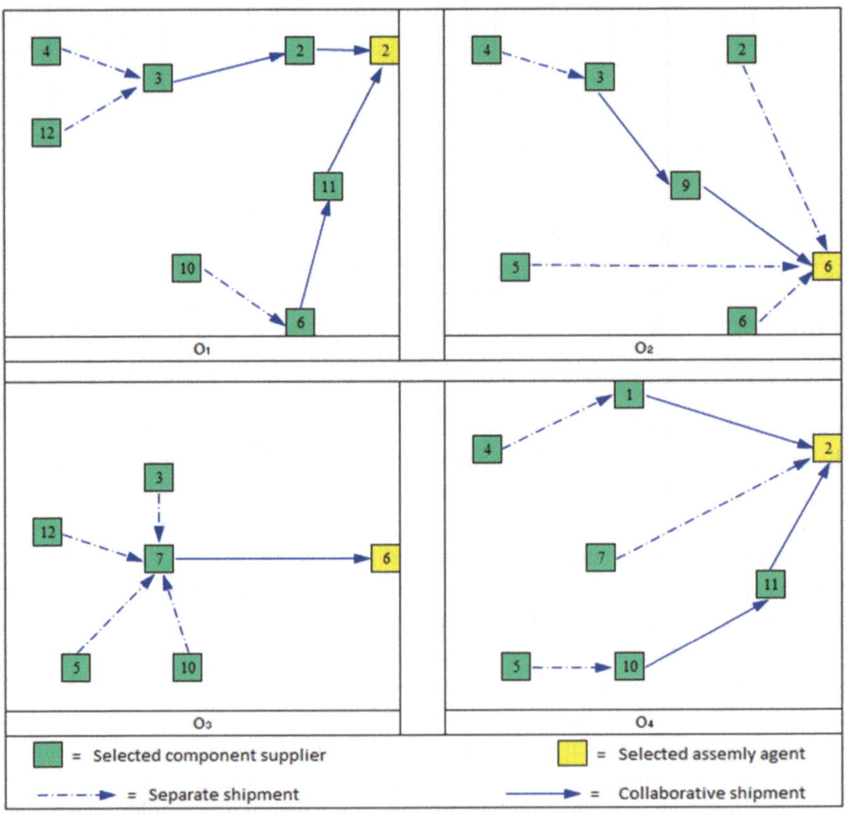

Fig. 4.1 Visualisation of the component shipping schedule

Complete component shipping schedule is the combination of all individual component shipping schedules of the product orders under consideration. When combining the schedules, the assumption of *maximum one shipment from one place to another* in one planning horizon is applied. Therefore, when combining the schedules if there are multiple shipments from one place to another, the multiple shipments are merged as one shipment. For example, two shipments from component supplier 4 to component supplier 3 in the schedules O_1 and O_2 or two shipments from component supplier 11 to assembly agent 2 in the schedules O_1 and O_4, as shown in Fig. 4.1, will be merged as one. Thereby, the shipping cost can be reduced.

The final part of a chromosome is about shipment scheduling to transport the products from the selected assembly agents to the customers, as illustrated in Table 4.3. The first column represents the assembly agents that were selected in the second part of a chromosome, for example assembly agents 2 and 6. The decision variables are in the highlighted cells. Tier 1 represents the corresponding customers who receive their products through direct shipments from the corresponding assembly agents. In other words, there is a shipment from the corresponding assembly agent in the first column to the corresponding customer in Tier 1, for instance from assembly agent 2 to customer 4. There may be a number of Tiers in the fourth part of a chromosome, depending on the number of customers involved in the assembly agents. In the illustrated example in Table 4.1, each assembly agent 2 or 6 is selected to assemble the products for two customers. Therefore, the corresponding fourth part of a chromosome has two Tiers as shown in Table 4.3. Each Tier, except the first one, has two columns that represent (1) the customers highlighted in yellow and (2) the product shipping decisions highlighted in green; but Tier 1 has only one column representing the customers only; the related product shipping decisions are decoded as explained above. In addition, all customers in each row in the fourth part of a chromosome must be those whose products are all assembled in the same assembly agent. For instance, in Table 4.3, the products requested by customers 1 and 4 are all assembled in assembly agent 2.

The product shipping decisions in green cells in Table 4.3 are binary numbers and interpreted as follows: there is a shipment from the corresponding customer in one Tier to the corresponding customer in the next Tier if the binary number representing the corresponding product shipping decision in the next Tier is 1. For example, there is a shipment from customer 4 to customer 1, as shown in Table 4.3.

Table 4.3 Fourth part of a chromosome—product shipment scheduling

Assembly agent	Tier 1	Tier 2	
2	4	1	1
6	3	0	2

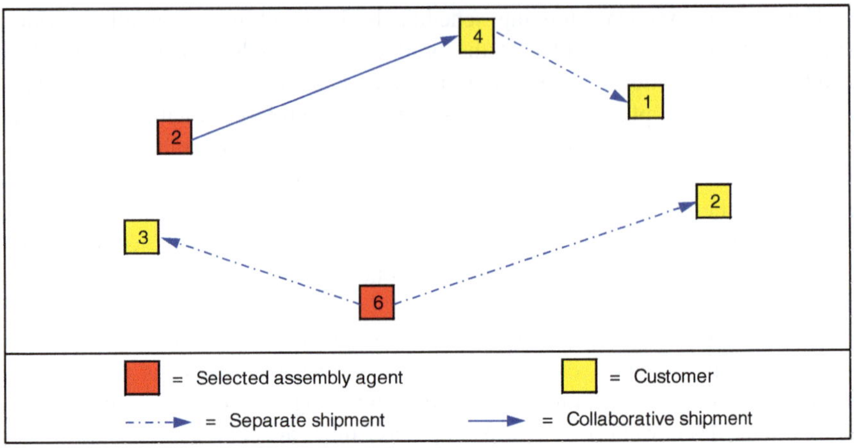

Fig. 4.2 Visualisation of the product shipping schedule

If the corresponding binary number in a Tier is 0, there is a shipment from the corresponding assembly agent to the corresponding customer in that Tier. For instance, there is a shipment from assembly agent 6 to customer 2, as shown in Table 4.3. The product shipping schedule encoded in the fourth part of a chromosome in Table 4.3 is visualised in Fig. 4.2.

It should be noted that the size of the fourth part of a chromosome is determined by the second part, because both number of rows (number of assembly agents involved) and number of columns (2 × Tiers—1) are determined by the selection in the second part of a chromosome. In other words, the fourth part of a chromosome is partially connected to the second part. In addition, the size of the fourth part of a chromosome can be changed from chromosome to chromosome, since the number of assembly agents as well as the number of Tiers involved may be different. Moreover, number of the active Tiers related to different assembly agent may be not the same, because number of products assembled in different assembly agent may be different. To illustrate the dynamic size of the fourth part of a chromosome, a number of possible scenarios are shown in Table 4.4.

In order to implement the proposed GA in computer software, the automatic generation of chromosomes, i.e. complete solutions to the problem, is required. Due to the complex constraints involved, generating a feasible chromosome as described above is not a trivial task and the following procedure is therefore proposed:

Table 4.4 Dynamic size of the fourth part of a chromosome

Fourth part of a chromosome								Decoded product shipping route
Assembly agent	Tier 1	Tier 2		Tier 3				1, A2 → Cus4 → Cus1 → Cus2 2, A6 → Cus3
2	4	1	1	1	2			
6	3	-	-	-	-			
Assembly agent	Tier 1	Tier 2		Tier 3				1, A2 → Cus4 → Cus1 2, A2 → Cus2 3, A6 → Cus3
2	4	1	1	0	2			
6	3	-	-	-	-			
Assembly agent	Tier 1	Tier 2		Tier 3				1, A2 → Cus4 2, A2 → Cus1 → Cus2 3, A6 → Cus3
2	4	0	1	1	2			
6	3	-	-	-	-			
Assembly agent	Tier 1	Tier 2		Tier 3		Tier 4		1, A2 → Cus4 → Cus1 → Cus2 → Cus3
2	4	1	1	1	2	1	3	
Assembly agent	Tier 1	Tier 2		Tier 3		Tier 4		1, A2 → Cus4 → Cus1 2, A2 → Cus2 → Cus3
2	4	1	1	0	2	1	3	
Assembly agent	Tier 1	Tier 2						1, A2 → Cus4 → Cus1 2, A6 → Cus3 3, A3 → Cus2
2	4	1	1					
6	3	-	-					
3	2	-	-					
Assembly agent	Tier 1	Tier 2						1, A2 → Cus4 2, A2 → Cus1 3, A6 → Cus3 4, A3 → Cus2
2	4	0	1					
6	3	-	-					
3	2	-	-					
Assembly agent	Tier 1							1, A2 → Cus4 2, A6 → Cus3 3, A3 → Cus2 4, A5 → Cus1
2	4							
6	3							
3	2							
5	1							

Note: A = Assembly agent, Cus = Customer

Step 1:	Randomly select one product order
Step 2:	Randomly select one component of the selected product order in Step 1
Step 3:	Randomly select one component supplier which is capable of providing the selected component in Step 2
Step 4:	Repeat Steps 2–3 for the rest of the components of the selected product order in Step 1
Step 5:	Randomly select one assembly agent which is capable of doing the final assembly operations for the selected product order in Step 1
Step 6:	Create a list of component suppliers selected in Step 3
Step 7:	Generate a provisional matrix that will represent the component shipping schedule of the selected product order in Step 1. It is noted that the provisional matrix is a null matrix but in the form of matrices O_1, O_2 … or O_4 in Table 4.2
Step 8:	Randomly select one component supplier from the list in Step 6
Step 9:	Randomly generate a binary number (0 or 1)

(continued)

(continued)

Step 10:	If the binary number in Step 9 is 0, assign a shipment from the component supplier selected in Step 8 to the assembly agent selected in Step 5 by setting the value of the corresponding cell of the matrix in Step 7 to 1; and then remove the selected component supplier in Step 8 from the list in Step 6 and finally go to Step 13
Step 11:	If the binary number in Step 9 is 1, randomly select another component supplier from the list in Step 6
Step 12:	Assign a shipment from component supplier selected in Step 8 to the one selected in Step 11 by setting the value of the corresponding cell of the matrix in Step 7 to 1 and then remove the selected component supplier in Step 8 from the list in Step 6
Step 13:	Repeat Steps 8–12 for the rest of the component suppliers, except the last one, in the list in Step 6
Step 14:	Assign a shipment from the last component supplier in Step 6 to the assembly agent selected in Step 5 by setting the value of the corresponding cell of the matrix in Step 7 to 1
Step 15:	Repeat Steps 1–14 for the rest of product orders
Step 16:	Make a list of assembly agents selected in Step 5
Step 17:	Randomly select one assembly agent from the list in Step 16
Step 18:	Randomly allocate the customers, whose products are assembled in the selected assembly agent in Step 17, to the Tiers as described in Tables 4.3 and 4.4
Step 19:	Randomly generate the product shipping decisions for the products assembled in the selected assembly agent in Step 17, as described in Tables 4.3 and 4.4
Step 20:	Repeat Steps 17–19 for the rest of the assembly agents in the list in Step 16

It can be seen that due to the nature of the VCIM production scheduling problem, such unique chromosome structure shown above is required to encode the solution. In addition, because of the unique chromosome structure, crossover and mutation of the proposed GA must be customised. The proposed crossover and mutation will be explained in the next Sections.

4.3 Crossover

Due to the unique chromosome structure as described in Sect. 4.2, crossover of the proposed GA must be customised; three crossover operations applied to three different parts of chromosomes, namely crossover 1, crossover 2 and crossover 3, are therefore proposed herein. The three operations are needed to ensure the thorough exchange of information between solutions in the evolution process.

4.3.1 Crossover 1

The first crossover operation is applied to the first part of chromosomes, i.e. component supplier selection part, as illustrated in Table 4.5. Although crossover 1

Table 4.5 Crossover 1

Input

	C1	C2	C3	C4	C5	C6	C7	C8	C9	C10
O1	4	12	3	2	11	6	10	2		
O2	6	5	9	3	2	4	4			
O3	3	12	5	10	7	5	3	5	3	7
O4	4	1	7	11	7	10	10	5	4	

	C1	C2	C3	C4	C5	C6	C7	C8	C9	C10
O1	7	11	1	2	3	9	8	12		
O2	1	7	11	2	3	6	10			
O3	10	11	2	12	9	6	6	6	2	5
O4	5	6	12	1	1	2	7	8	9	

Output

	C1	C2	C3	C4	C5	C6	C7	C8	C9	C10
O1	4	12	3	2	3	9	8	12		
O2	6	5	9	3	3	6	10			
O3	3	12	5	10	9	6	6	6	2	5
O4	4	1	7	11	1	2	7	8	9	

	C1	C2	C3	C4	C5	C6	C7	C8	C9	C10
O1	7	11	1	2	11	6	10	2		
O2	1	7	11	2	2	4	4			
O3	10	11	2	12	7	5	3	5	3	7
O4	5	6	12	1	7	10	10	5	4	

is applied only to part 1 of chromosomes, it has an effect on the corresponding part 3, component shipment scheduling part, because the two parts are interconnected as mentioned in Sect. 4.2. In addition, the offspring chromosomes may need to be repaired to ensure their feasibilities because some constraints may be violated due to the crossover operation. To implement the crossover, the following steps are proposed.

Step 1:	Randomly select two parent chromosomes
Step 2:	Determine the component supplier selection part (part 1) of the selected parent chromosomes in Step 1, the input as shown in Table 4.5
Step 3:	Randomly select a cut point and swap the pieces as illustrated in Table 4.5
Step 4:	Randomly select one offspring chromosome
Step 5:	Randomly select one product order encoded in the offspring selected in Step 4
Step 6:	Determine the component suppliers which are no longer involved in the order selected in Step 5, because of the crossover operation
Step 7:	Remove the component suppliers determined in Step 6 from the corresponding component shipping schedule in part 3 of the offspring chromosome and repair the associated shipment tours by connecting their previous adjacent component suppliers to the related assembly agent
Step 8:	Determine the new component suppliers emerged due to the crossover operation
Step 9:	Randomly add the component suppliers determined in Step 8 to the updated component shipping schedule in Step 7
Step 10:	Repeat Steps 5–9 for the rest of product orders encoded in the offspring selected in Step 4
Step 11:	Repeat Steps 4–10 for the rest of offspring chromosomes
Step 12:	Form the offspring chromosomes accordingly

It can be seen that Steps 1–3 are for "actual" crossover operation and Steps 4–11 are for repairing part 3 of offspring chromosomes, as part 3 may not be feasible after the crossover operation. There is no need to repair part 1 because it is always feasible. In addition, parts 2 and 4 of offspring chromosomes are not affected by crossover 1.

Table 4.6 Crossover 2

Input

	A
O1	2
O2	6
O3	6
O4	2

	A
O1	4
O2	6
O3	3
O4	4

Output

	A
O1	2
O2	6
O3	3
O4	4

	A
O1	4
O2	6
O3	6
O4	2

4.3.2 *Crossover 2*

The second crossover operation is applied to part 2 of parent chromosomes, assembly agent selection part, as illustrated in Table 4.6. Although crossover 2 is applied only to part 2 of chromosomes it has an effect on the corresponding part 4, product shipment scheduling part, because the two parts are interconnected as mentioned in Sect. 4.2. In addition, the offspring chromosomes may need to be repaired to ensure their feasibilities since some constraints may be violated after the crossover operation. The following steps are proposed to implement crossover 2.

Step 1:	Randomly select two parent chromosomes
Step 2:	Determine the assembly agent selection part (part 2) of the selected parent chromosomes in Step 1, the input as shown in Table 4.6
Step 3:	Randomly select a cut point and swap the pieces as illustrated in Table 4.6
Step 4:	Randomly select one offspring chromosome
Step 5:	Replace part 4 of the selected offspring chromosome in Step 4 by a new one which is randomly generated using Steps 6–10
Step 6:	Make a list of assembly agents in part 2 of the selected offspring chromosome in Step 4
Step 7:	Randomly select one assembly agent from the list in Step 6
Step 8:	Randomly allocate the customers, whose products are assembled in the selected assembly agent in Step 7, to the Tiers as described in Tables 4.3 and 4.4
Step 9:	Randomly generate the product shipping decisions for the products assembled in the selected assembly agent in Step 8, as described in Table 4.3 and 4.4
Step 10:	Repeat Steps 7–9 for the rest of assembly agents in the list in Step 6
Step 11:	Repeat Steps 4–10 for the rest of offspring chromosomes
Step 12:	Form the offspring chromosomes accordingly

It should be noted that both parts 2 and 4 of offspring chromosomes are affected by crossover 2. Part 2 is always feasible after the crossover operation and part 4 is guaranteed to be feasible by Steps 4–11. It can be seen that part 4 of offspring chromosomes is not repaired but randomly regenerated; there are two reasons behind this strategy. First, repairing part 4 of offspring chromosomes is very complex as many constraints are involved. Second, randomly regenerating part 4 of offspring chromosomes is more efficient than repairing it.

4.3.3 Crossover 3

The third crossover operation is applied to part 3, component shipment scheduling part, of chromosomes. To explore solution space of the component shipment scheduling part, the crossover operation is proposed to be implemented by the following steps.

Step 1:	Randomly select one parent chromosome
Step 2:	Randomly select one component shipping schedule in part 3 of the selected parent chromosome in Step 1
Step 3:	Replace the selected component shipping schedule in Step 2 by a new one which is randomly generated using Steps 4–12
Step 4:	Retain the list of component suppliers involved in the selected component shipping schedule in Step 2
Step 5:	Delete all decision variables in the matrix representing the selected component shipping schedule in Step 2
Step 6:	Randomly select one component supplier from the list in Step 4
Step 7:	Randomly generate a binary number (0 or 1)
Step 8:	If the binary number in Step 7 is 0, assign a shipment from the component supplier selected in Step 6 to the related assembly agent by setting the value of the corresponding cell of the matrix in Step 5 to 1; and then remove the selected component supplier in Step 6 from the list in Step 4 and finally go to Step 11
Step 9:	If the binary number in Step 7 is 1, randomly select another component supplier from the list in Step 4
Step 10:	Assign a shipment from component supplier selected in Step 6 to the one selected in Step 9 by setting the value of the corresponding cell of the matrix in Step 5 to 1 and then remove the selected component supplier in Step 6 from the list in Step 4
Step 11:	Repeat Steps 6–10 for the rest of the component suppliers, except the last one, in the list in Step 4
Step 12:	Assign a shipment from the last component supplier in Step 4 to the related assembly agent by setting the value of the corresponding cell of the matrix in Step 5 to 1. It is noted that the matrix in Step 5 now represents the new component shipping schedule
Step 13:	Form the offspring chromosome accordingly

As can be seen from Steps 1–13, unlike crossovers 1–2, crossover 3 has only one parent chromosome as the input and has only one offspring chromosome as the output. In addition, crossover 3 does not have gene swapping operations like the ones in Tables 4.5 and 4.6, because such swapping operations in crossover 3 involve many complex constraints that make the crossover inefficient. Moreover, crossover 3 does not affect parts 1, 2 and 4 of chromosomes.

4.4 Mutation

Due to the unique chromosome structure as described in Sect. 4.2, mutation of the proposed GA must be customised; four mutation operations applied to four different parts of chromosomes, namely mutation 1, mutation 2, mutation 3, and mutation 4, are therefore proposed herein.

4.4.1 Mutation 1

The first mutation operation is applied to the first part of chromosomes, component supplier selection part, as illustrated in Table 4.7. Although mutation 1 is applied only to part 1 of chromosomes it has an effect on the corresponding part 3, component shipment scheduling part, because the two parts are interconnected as mentioned in Sect. 4.2. In addition, the offspring chromosomes may need to be repaired to ensure their feasibilities because some constraints may be violated due to the mutation operation. To implement the mutation, the following steps are proposed.

Table 4.7 Mutation 1

Input

	C_1	C_2	C_3	C_4	C_5	C_6	C_7	C_8	C_9	C_{10}
O_1	4	12	3	2	11	6	10	2		
O_2	6	5	9	3	2	4	4			
O_3	3	12	5	10	7	5	3	5	3	7
O_4	4	1	7	11	7	10	10	5	4	

	C_1	C_2	C_3	C_4	C_5	C_6	C_7	C_8	C_9	C_{10}
O_1	7	11	1	2	3	9	8	12		
O_2	1	7	11	2	3	6	10			
O_3	10	11	2	12	9	6	6	6	2	5
O_4	5	6	12	1	1	2	7	8	9	

Output*

	C_1	C_2	C_3	C_4	C_5	C_6	C_7	C_8	C_9	C_{10}
O_1	4	12	3	2	11	6	10	2		
O_2	6	6	9	3	2	4	4			
O_3	3	12	5	10	7	5	3	5	3	7
O_4	4	1	7	11	7	10	10	5	4	

	C_1	C_2	C_3	C_4	C_5	C_6	C_7	C_8	C_9	C_{10}
O_1	7	11	1	2	3	9	8	12		
O_2	1	7	11	2	3	6	10			
O_3	10	11	2	12	9	6	5	6	2	5
O_4	5	6	12	1	1	2	7	8	9	

* *These parts of the offspring chromosomes may need to be repaired to ensure their feasibilities.*

Step 1:	Randomly select two parent chromosomes
Step 2:	Determine the component supplier selection part (part 1) of the selected parent chromosomes in Step 1, the input as shown in Table 4.7
Step 3:	Randomly select two genes and swap them as illustrated in Table 4.7
Step 4:	Replace infeasible values in part 1 of the offspring chromosomes by feasible ones which are randomly generated
Step 5:	Randomly select one offspring chromosome
Step 6:	Randomly select one product order encoded in the offspring selected in Step 5
Step 7:	Determine the component suppliers which are no longer involved in the order selected in Step 6, because of the crossover operation
Step 8:	Remove the component suppliers determined in Step 7 from the corresponding component shipping schedule in part 3 of the offspring chromosome and repair the associated shipment tours by connecting their previous adjacent component suppliers to the related assembly agent
Step 9:	Determine the new component suppliers emerged due to the crossover operation
Step 10:	Randomly add the component suppliers determined in Step 9 to the updated component shipping schedule in Step 8
Step 11:	Repeat Steps 6–10 for the rest of product orders encoded in the offspring selected in Step 5
Step 12:	Repeat Steps 5–11 for the rest of offspring chromosomes
Step 13:	Form the offspring chromosomes accordingly

It should be noted that Steps 1–3 are for "actual" mutation operation, Step 4 is for repairing part 1, and Steps 5–12 are for repairing part 3 of offspring chromosomes; the repairs are required to make sure the offspring chromosomes are feasible. In addition, parts 2 and 4 of offspring chromosomes are not affected by mutation 1.

4.4.2 *Mutation 2*

The second mutation operation is applied to the second part of chromosomes, assembly agent selection part, as illustrated in Table 4.8. Although mutation 2 is

Table 4.8 Mutation 2

Input

	A
O1	2
O2	6
O3	6
O4	2

	A
O1	4
O2	6
O3	3
O4	4

Output*

	A
O1	3
O2	6
O3	6
O4	2

	A
O1	4
O2	6
O3	2
O4	4

* *These parts of the offspring chromosomes may need to be repaired to ensure their feasibilities.*

applied only to part 2 of chromosomes it has an effect on the corresponding part 4, product shipment scheduling part, because the two parts are interconnected as mentioned in Sect. 4.2. In addition, the offspring chromosomes may need to be repaired to ensure their feasibilities since some constraints may be violated because of the mutation operation. To implement mutation 2, the following steps are proposed.

Step 1:	Randomly select two parent chromosomes
Step 2:	Determine the assembly agent selection part (part 2) of the selected parent chromosomes in Step 1, the input as shown in Table 4.8
Step 3:	Randomly select two genes and swap them as illustrated in Table 4.8
Step 4:	Replace infeasible values in part 2 of the offspring chromosomes by feasible ones which are randomly generated
Step 5:	Randomly select one offspring chromosome
Step 6:	Replace part 4 of the selected offspring chromosome in Step 5 by a new one which is randomly generated using Steps 7–11
Step 7:	Make a list of assembly agents in part 2 of the selected offspring chromosome in Step 5
Step 8:	Randomly select one assembly agent from the list in Step 7
Step 9:	Randomly allocate the customers, whose products are assembled in the selected assembly agent in Step 8, to the Tiers as described in Tables 4.3 and 4.4
Step 10:	Randomly generate the product shipping decisions for the products assembled in the selected assembly agent in Step 8, as described in Tables 4.3 and 4.4
Step 11:	Repeat Steps 8–10 for the rest of assembly agents in the list in Step 7
Step 12:	Repeat Steps 5–11 for the rest of offspring chromosomes
Step 13:	Form the offspring chromosomes accordingly

It should be noted that both parts 2 and 4 of offspring chromosomes are affected by mutation 2. Part 2 is repaired to ensure its feasibility by Step 4 and part 4 is guaranteed to be feasible by Steps 5–12. As can be seen, part 4 of offspring chromosomes is not repaired but randomly regenerated; there are two reasons behind this strategy. First, repairing part 4 of offspring chromosomes is very complex as many constraints involved. Second, randomly regenerating part 4 of offspring chromosomes is more efficient than repairing it.

4.4.3 Mutation 3

The third mutation operation is applied to part 3, component shipment scheduling part, of chromosomes, as illustrated in Tables 4.9, 4.10, 4.11, 4.12, 4.13, 4.14. Due to unique structure and complex constraints of the part, the following steps are proposed to implement mutation 3.

Table 4.9 Mutation 3—Example 1

Component shipping schedule in chromosome form

Input

	S1	S2	S3	S4	S5	S6	S7	S8	S9	S10	S11	S12	A
S1			1										
S2													
S3						1							
S4								1					
S5													
S6													1
S7													
S8													1
S9													
S10													
S11								1					
S12													

Output

	S1	S2	S3	S4	S5	S6	S7	S8	S9	S10	S11	S12	A
S1			1										
S2													
S3						1							
S4								1					
S5													
S6								1					
S7													
S8													1
S9													
S10													
S11								1					
S12													

Visualisation of component shipping schedule

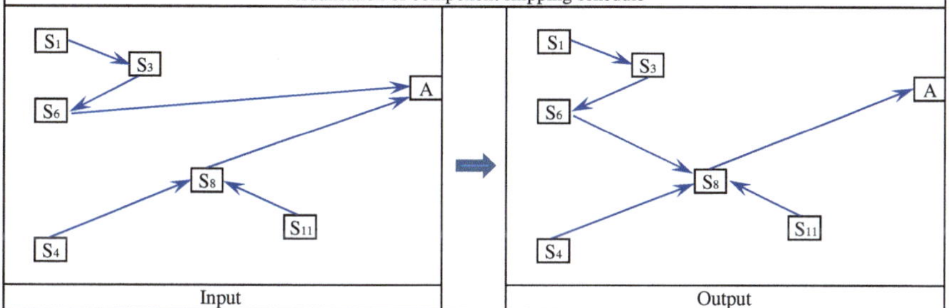

Si = Selected manufacturing agent → = Shipment A = Selected assembly agent

Table 4.10 Mutation 3—Example 2

Component shipping schedule in chromosome form

Input

	S1	S2	S3	S4	S5	S6	S7	S8	S9	S10	S11	S12	A
S1			1										
S2													
S3						1							
S4								1					
S5													
S6													1
S7													
S8													1
S9													
S10													
S11								1					
S12													

Output

	S1	S2	S3	S4	S5	S6	S7	S8	S9	S10	S11	S12	A
S1													1
S2													
S3						1							
S4								1					
S5													
S6	1												
S7													
S8													1
S9													
S10													
S11								1					
S12													

Visualisation of component shipping schedule

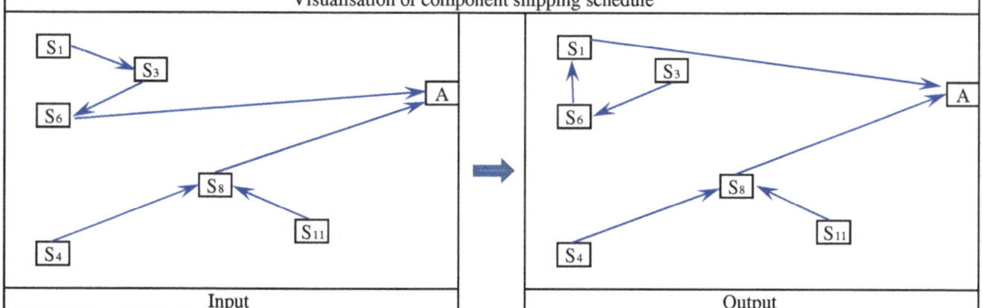

Si = Selected manufacturing agent → = Shipment A = Selected assembly agent

Table 4.11 Mutation 3—Example 3

Component shipping schedule in chromosome form

Input

	S1	S2	S3	S4	S5	S6	S7	S8	S9	S10	S11	S12	A
S1			1										
S2													
S3						1							
S4								1					
S5													
S6													1
S7													
S8													1
S9													
S10													
S11								1					
S12													

Output

	S1	S2	S3	S4	S5	S6	S7	S8	S9	S10	S11	S12	A
S1			1										
S2													
S3													1
S4								1					
S5													
S6			1										
S7													
S8													1
S9													
S10													
S11								1					
S12													

Visualisation of component shipping schedule

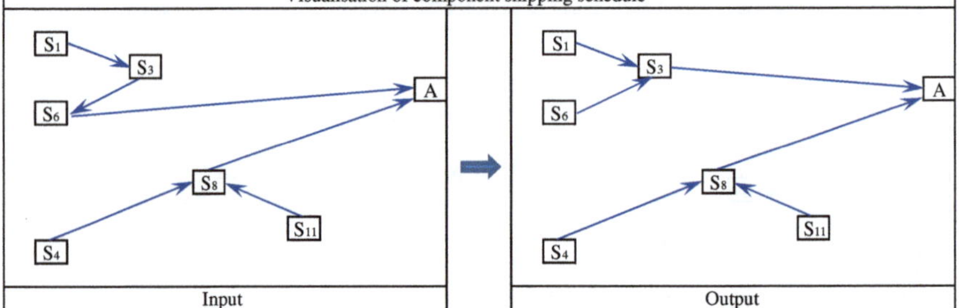

Si = Selected manufacturing agent; → = Shipment; A = Selected assembly agent

Table 4.12 Mutation 3—Example 4

Component shipping schedule in chromosome form

Input

	S1	S2	S3	S4	S5	S6	S7	S8	S9	S10	S11	S12	A
S1			1										
S2													
S3						1							
S4								1					
S5													
S6													1
S7													
S8													1
S9													
S10													
S11								1					
S12													

Output

	S1	S2	S3	S4	S5	S6	S7	S8	S9	S10	S11	S12	A
S1			1										
S2													
S3						1							
S4													1
S5													
S6				1									
S7													
S8													1
S9													
S10													
S11								1					
S12													

Visualisation of component shipping schedule

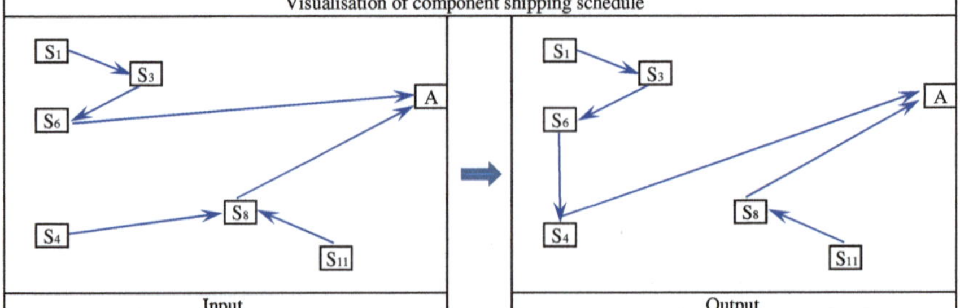

Si = Selected manufacturing agent; → = Shipment; A = Selected assembly agent

Table 4.13 Mutation 3—Example 5

Component shipping schedule in chromosome form

Input

	S1	S2	S3	S4	S5	S6	S7	S8	S9	S10	S11	S12	A
S1			1										
S2													
S3						1							
S4								1					
S5													
S6													1
S7													
S8													1
S9													
S10													
S11								1					
S12													

Output

	S1	S2	S3	S4	S5	S6	S7	S8	S9	S10	S11	S12	A
S1			1										
S2													
S3						1							
S4						1							
S5													
S6													1
S7													
S8													1
S9													
S10													
S11								1					
S12													

Visualisation of component shipping schedule

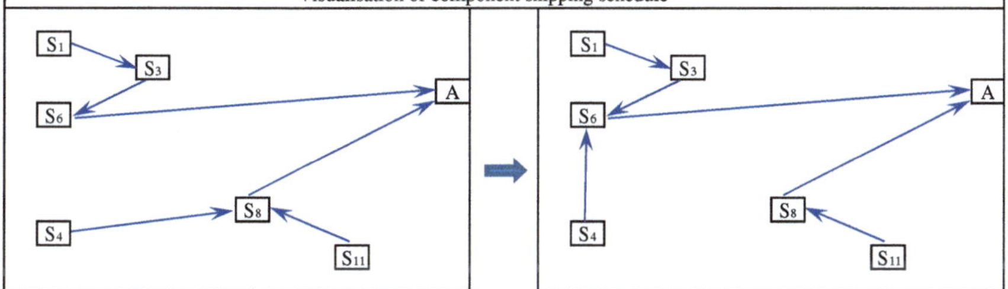

Input / Output

Si = Selected manufacturing agent → = Shipment A = Selected assembly agent

Table 4.14 Mutation 3—Example 6

Component shipping schedule in chromosome form

Input

	S1	S2	S3	S4	S5	S6	S7	S8	S9	S10	S11	S12	A
S1			1										
S2													
S3						1							
S4								1					
S5													
S6													1
S7													
S8													1
S9													
S10													
S11								1					
S12													

Output

	S1	S2	S3	S4	S5	S6	S7	S8	S9	S10	S11	S12	A
S1			1										
S2													
S3						1							
S4													1
S5													
S6													1
S7													
S8													1
S9													
S10													
S11								1					
S12													

Visualisation of component shipping schedule

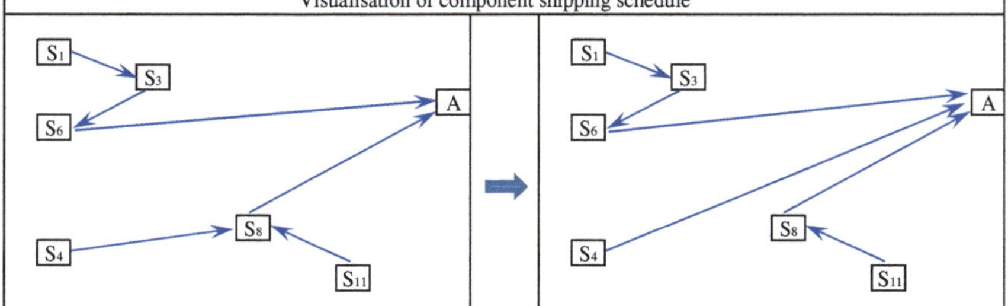

Input / Output

Si = Selected manufacturing agent → = Shipment A = Selected assembly agent

Step 1:	Randomly select one parent chromosome
Step 2:	Randomly select one component shipping schedule in part 3 of the selected parent chromosome in Step 1
Step 3:	Randomly select one gene with the value of 1
Step 4:	If the selected gene in Step 3 is in column A (the yellow gene in the first sub-table in Table 4.9, for example), go to Step 5; otherwise, go to Step 8
Step 5:	Randomly select another gene with the value of 0 satisfying two following conditions: (1) being in the same row as the selected gene in Step 3, and (2) being in a feasible column (a gene that is highlighted in red in the first sub-table in Table 4.9, for instance). It is noted that feasible column herein is referred to a column corresponding to a component supplier that is involved in the component shipping schedule, for example, columns S_1, S_3, S_4, S_6, S_8 and S_{11} in the first sub-table in Table 4.9
Step 6:	Exchange the two selected genes in Steps 3 and 5 as illustrated in Tables 4.9, 4.10, 4.11, 4.12
Step 7:	Repair the shipment tour associated with the component supplier in the feasible column mentioned in Step 5 by assigning a new shipment that goes directly from that component supplier to the assembly agent as illustrated in Tables 4.9, 4.10, 4.11, 4.12; and then go to Step 10. It should be noted that the changes in the component shipping schedule because of the mutation are visualised in the accompanying figures in Tables 4.9, 4.10, 4.11, 4.12
Step 8:	Randomly select another gene with the value of 0 satisfying two following conditions: (1) being in the same row as the selected gene in Step 3, and (2) being in the column A or a feasible column (for example, a gene that is highlighted in yellow in the first sub-table in Tables 4.13 or 4.14)
Step 9:	Exchange the two selected genes in Steps 3 and 8 as illustrated in Tables 4.13 and 4.14. Again, the changes in the component shipping schedules due to the mutation are visualised in the accompanying figures in Tables 4.13 and 4.14
Step 10:	Form the offspring chromosome accordingly

As can be seen from Steps 1–10, like crossover 3, mutation 3 has only one parent chromosome as the input and has only one offspring chromosome as the output. In addition, parts 1, 2 and 4 of chromosomes are not affected by mutation 3.

4.4.4 *Mutation 4*

The fourth mutation operation is applied to part 4, product shipment scheduling part, of chromosomes, as illustrated in Tables 4.15, 4.16, 4.17, 4.18, 4.19, 4.20, 4.21. Due to the unique structure of the product shipment scheduling part, the following steps are proposed to implement mutation 4.

Table 4.15 Mutation 4—Example 1

Product shipping schedule in chromosome form

Input

Assembly agent	Tier 1	Tier 2	
2	4	1	1
6	3	1	2

Output

Assembly agent	Tier 1	Tier 2	
2	4	0	1
6	3	1	2

Decoded product shipping route

Input

Route 1:	A2 → Cus4 → Cus1
Route 2:	A6 → Cus3 → Cus2

Output

Route 1:	A2 → Cus4
Route 2:	A2 → Cus1
Route 3:	A6 → Cus3 → Cus2

Note: A = Assembly agent, Cus = Customer

Table 4.16 Mutation 4—Example 2

Product shipping schedule in chromosome form

Input

Assembly agent	Tier 1	Tier 2		Tier 3	
2	4	1	1	1	2
6	3	-	-	-	-

Output

Assembly agent	Tier 1	Tier 2		Tier 3	
2	4	0	1	1	2
6	3	-	-	-	-

Decoded product shipping route

Input

Route 1:	A2 → Cus4 → Cus1 → Cus2
Route 2:	A6 → Cus3

Output

Route 1:	A2 → Cus4
Route 2:	A2 → Cus1 → Cus2
Route 3:	A6 → Cus3

Note: A = Assembly agent, Cus = Customer

Table 4.17 Mutation 4—Example 3

Product shipping schedule in chromosome form

Input

Assembly agent	Tier 1	Tier 2		Tier 3	
2	4	1	1	0	2
6	3	-	-	-	-

Output

Assembly agent	Tier 1	Tier 2		Tier 3	
2	4	1	1	1	2
6	3	-	-	-	-

Decoded product shipping route

Input

Route 1:	A2 → Cus4 → Cus1
Route 2:	A2 → Cus2
Route 3:	A6 → Cus3

Output

Route 1:	A2 → Cus4 → Cus1 → Cus2
Route 2:	A6 → Cus3

Note: A = Assembly agent, Cus = Customer

Table 4.18 Mutation 4—Example 4

Product shipping schedule in chromosome form

Assembly agent	Tier 1	Tier 2		Tier 3		→	Assembly agent	Tier 1	Tier 2		Tier 3	
2	4	1	1	1	2		2	2	1	1	1	4
6	3	-	-	-	-		6	3	-	-	-	-
Input							Output					

Decoded product shipping route

Input	→	Output
Route 1: A2 → Cus4 → Cus1 → Cus2		Route 1: A2 → Cus2 → Cus1 → Cus4
Route 2: A6 → Cus3		Route 2: A6 → Cus3

Note: A = Assembly agent, Cus = Customer

Table 4.19 Mutation 4—Example 5

Product shipping schedule in chromosome form

Assembly agent	Tier 1	Tier 2		Tier 3		→	Assembly agent	Tier 1	Tier 2		Tier 3	
2	4	1	1	1	2		2	4	1	2	1	1
6	3	-	-	-	-		6	3	-	-	-	-
Input							Output					

Decoded product shipping route

Input	→	Output
Route 1: A2 → Cus4 → Cus1 → Cus2		Route 1: A2 → Cus4 → Cus2 → Cus1
Route 2: A6 → Cus3		Route 2: A6 → Cus3

Note: A = Assembly agent, Cus = Customer

Table 4.20 Mutation 4—Example 6

Product shipping schedule in chromosome form

Assembly agent	Tier 1	Tier 2		Tier 3		→	Assembly agent	Tier 1	Tier 2		Tier 3	
2	4	1	1	1	2		2	1	1	4	1	2
6	3	-	-	-	-		6	3	-	-	-	-
Input							Output					

Decoded product shipping route

Input	→	Output
Route 1: A2 → Cus4 → Cus1 → Cus2		Route 1: A2 → Cus1 → Cus4 → Cus2
Route 2: A6 → Cus3		Route 2: A6 → Cus3

Note: A = Assembly agent, Cus = Customer

Table 4.21 Mutation 4—Example 7

Product shipping schedule in chromosome form

Input

Assembly agent	Tier 1	Tier 2	
2	4	1	1
6	3	1	2

Output

Assembly agent	Tier 1	Tier 2	
2	4	1	1
6	2	1	3

Decoded product shipping route

Input

Route 1: A2 → Cus4 → Cus1
Route 2: A6 → Cus3 → Cus2

Output

Route 1: A2 → Cus4 → Cus1
Route 2: A6 → Cus2 → Cus3

Note: A = Assembly agent, Cus = Customer

Step 1:	Randomly select one parent chromosome
Step 2:	Determine the number of Tiers in part 4 of the selected parent chromosome in Step 1
Step 3:	If the number of Tiers in Step 2 is greater than one, go to Step 4; otherwise, adopt part 4 of the parent chromosome as part 4 of the corresponding offspring chromosome, and then go to Step 9
Step 4:	Randomly select one assembly agent which assembles products for more than one customer, assembly agent 2 in Table 4.15, for example
Step 5:	Randomly select one gene in the row representing the selected assembly agent in Step 4
Step 6:	If the selected gene in Step 5 represents the product shipping decision, flip the gene value as illustrated in Tables 4.15, 4.16, 4.17, and then go to Step 9
Step 7:	If the selected gene in Step 5 represents the customer, randomly select another gene representing another customer
Step 8:	Exchange the two selected genes in Steps 5 and 7 as illustrated in Tables 4.18, 4.19, 4.20, 4.21
Step 9:	Form the offspring chromosome accordingly

As can be seen from Steps 1–9, mutation 4 has only one parent chromosome as the input and has only one offspring chromosome as the output. In addition, in a case where part 4 of parent chromosome has only one Tier, offspring chromosome is exactly the same as the parent chromosome; in other words, in this case, there is no mutation operation. Moreover, offspring chromosomes of mutation 4 are always feasible; therefore repairing the offspring chromosomes to ensure their feasibilities is not required.

4.5 Evaluation

As mentioned in Sect. 3.2, there are three objective functions in the proposed VCIM production scheduling model, i.e. minimum total cost, minimum average completion time and maximum average reliability. There are a number of approaches to solve multi-objective optimisation problems such as the weighted sum method (Marler and Arora 2009), the combined scalarizing method (Rastegar and Khorram 2014), goal programming-based method (Chen and Xu 2012), the ε-constraint method (Aghaei et al. 2011; Laumanns et al. 2006; Mavrotas 2009), the Pareto-based method (Giagkiozis and Fleming 2015), etc. Each approach has its own advantages and disadvantages. In this research, the ε-constraint method is chosen because of the following reasons. First, it is widely used to solve multi-objective optimisation (Chiandussi et al. 2012; Copado-Méndez et al. 2016; Rezvani et al. 2015). Second, it can handle non-convex Pareto sets (Copado-Méndez et al. 2016; Khalili-Damghani and Amiri 2012). Third, it allows the decision makers to set the threshold values for some objective functions (Chiandussi et al. 2012).

In the ε-constraint method, one objective function is optimised, while the rest of the objective functions are considered as constraints with the given threshold values (Cabrera et al. 2014; Kang et al. 2015; Kitayama and Yamazaki 2012; Li et al. 2015). In this research, the total cost objective function is considered to be minimised while two other objective functions, i.e. minimum average completion time and maximum average reliability are transformed into constraints with pre-set threshold values. Thereby, the decision makers can maintain a pre-set standard performance of their VCIM systems. It should be noted that an additional constraint handling strategy is required to deal with these two newly added constraints. In this research, all chromosomes in which these two constraints are not satisfied are replaced by new ones generated randomly using the proposed strategy in Sect. 4.2. This replacement is done inside the "Evaluation" box shown in Fig. 4.4.

It should be noted that the component release times, product assembly times as well as shipping times in the proposed VCIM production scheduling model in Chap. 3 are all stochastic parameters. Therefore Monte Carlo simulation is used to estimate the expected value of the average completion time of all product orders under consideration; as mentioned above, this value is the value of the second objective function and it must be smaller than or equal to the pre-set value to make the corresponding chromosome feasible. For detailed description of Monte Carlo simulation it is advised to refer to Brandimarte (2014). One of the issues in Monte Carlo simulation is the sample size which is herein estimated by a function named *Sample-Size-for-Estimation* in Minitab software.

Since the last two objective functions, i.e. minimum average completion time and maximum average reliability, are transformed into constraints, quality of chromosomes in this research is evaluated based on the value of the first objective function, i.e. the total cost of all product orders under consideration. Selecting chromosomes for the next generation, which is based on the total cost, will be presented in the next Section.

4.6 Selection

Selection is a driving force that directs GA search toward promising regions in the search space (Gen and Cheng 1997, p. 20). That is why the selection plays an important role in any GA. To date, a number of selection methods have been developed such as roulette-wheel selection (Zhao and Wang 2011), tournament selection (Castelli and Vanneschi 2014), probabilistic selection (Maity et al. 2015), linear ranking selection (Shopova and Vaklieva-Bancheva 2006), non-linear ranking selection (Ren et al. 2007), dynamic ranking selection (Boudissa and Bounekhla 2012), etc.

In this research, *dynamic ranking selection*, developed by Boudissa and Bounekhla (2012), is adopted because of the following reasons. First, the dynamic ranking selection varies the selection pressure along generation which allows GA to have a wider exploration in early generations and have more intense exploitation in late generations; this characteristic can enhance the capability of GA to find the global optimal solution. Second, the dynamic ranking selection is capable of preventing premature convergence problem caused by some super chromosomes, because it is not based on raw fitness values. Third, superiority of the dynamic ranking selection over other selection methods including roulette wheel selection, linear ranking selection and Boltzmann selection has been verified by Boudissa and Bounekhla (2012).

The dynamic ranking selection, developed by Boudissa and Bounekhla (2012), is described as follows. First, chromosomes in a generation are ranked based on the objective function values and then a dynamic function called quadratic fitness function is used to assign fitness values, based on the ranks, to the chromosomes. Next, the selection probability of a chromosome is calculated as shown in Eq. 4.1, where $p(i)$ is the selection probability of chromosome i, $Fit(i)$ is the fitness value of chromosome i, and N is the population size. Finally, the roulette-wheel procedure with the calculated selection probabilities is used to select chromosomes for the next generation. It should be noted that the quadratic fitness function is not a fixed function; it is changed along the generation of GA as visualised in Fig. 4.3. The quadratic fitness function is mathematically expressed by Eqs. 4.2–4.9, where $Fit(i)$ is the fitness value of ith chromosome, i is the rank of a chromosome, $Fmax$ is the maximum objective function value, $Fmin$ is the minimum objective function value, N is the population size, g is the current generation, and tg is the total generation.

$$p(i) = \frac{Fit(i)}{\sum_1^N Fit(i)} \tag{4.1}$$

$$Fit(i) = A(\rho)i^2 + B(\rho)i + C(\rho) \tag{4.2}$$

$$A(\rho) = \frac{(Fmax + Fmin)(1-\rho) + (N-\rho)Fmax + (\rho - 1)Fmin}{(1-N)\rho^2 + (\rho - 1)N^2 + N - \rho} \tag{4.3}$$

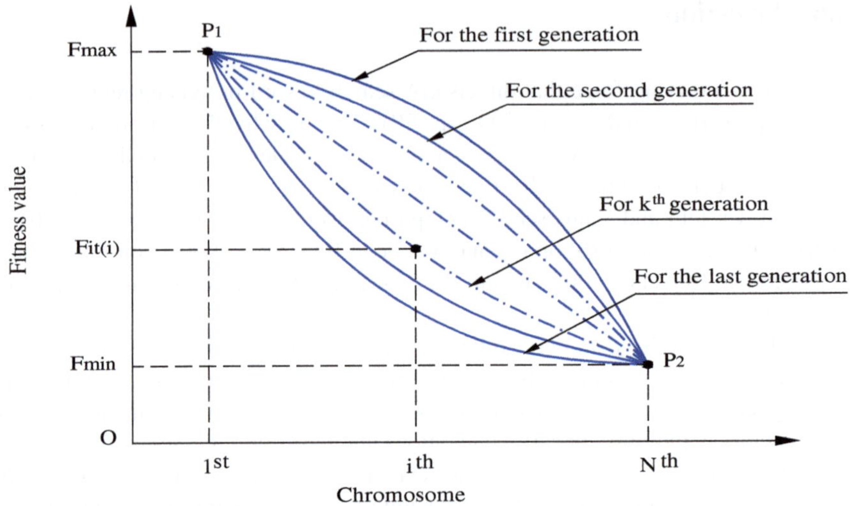

Note: Fit(i) is fitness value assigned to i th chromosome in k th generation.

Fig. 4.3 Quadratic fitness function [adapted from Boudissa and Bounekhla (2012)]

$$B(\rho) = \frac{(Fmax + Fmin)(\rho - 1)(N + 1) + (1 - \rho^2)Fmin + (\rho^2 - N^2)Fmax}{(1 - N)\rho^2 + (\rho - 1)N^2 + N - \rho} \quad (4.4)$$

$$C(\rho) = \frac{(Fmax + Fmin)(1 - \rho)N + (\rho - 1)\rho.Fmin + (N - \rho)\rho.N.Fmax}{(1 - N)\rho^2 + (\rho - 1)N^2 + N - \rho} \quad (4.5)$$

$$\rho = \rho_2 + (\rho_1 - \rho_2)\frac{g - 1}{tg - 1} \quad (4.6)$$

$$\rho_1 = N + \frac{(N - 1)(Fmax + Fmin) - (N - 1)^2\sqrt{\Delta}}{2(Fmax - Fmin)} \quad (4.7)$$

$$\rho_2 = 1 + \frac{(N - 1)^2\sqrt{\Delta} - (N - 1)(Fmax + Fmin)}{2(Fmax - Fmin)} \quad (4.8)$$

$$\Delta = \frac{(Fmax + Fmin)^2 + 4(Fmax - Fmin)Fmax}{(N - 1)^2} \quad (4.9)$$

4.7 Algorithm Structure and Parameter Tuning

To improve the global search capability, an innovative GA structure with an adaptive stop-and-restart-with-memory mechanism, as shown in Fig. 4.4, is proposed herein. With the developed structure, the proposed GA has a number of remarkable characteristics as follows. First, the proposed GA is capable of restarting its search process whenever the quality of the best chromosome obtained so far is not improved within a set number of generations. Second, its offspring chromosomes which are generated by three crossover and four mutation operators as described above, can compete with parent chromosomes for survival from one generation to the next. Third, a certain number of elite chromosomes are guaranteed to pass from one generation to the next. Finally, when the search process of the proposed GA is restarted, a certain number of elite chromosomes obtained so far are reserved and then put into the next initial population.

It is known that parameters such as population size, crossover rate, mutation rate, number of elite chromosomes, etc. have a significant effect on the efficiency and effectiveness of any GA. In this research, to maximise the efficiency and effectiveness of the proposed GA, Taguchi experimental design based tuning approach is developed to select its parameter set which includes population size (P), rate of crossover 1 (C_1), rate of crossover 2 (C_2), rate of crossover 3 (C_3), rate of mutation 1 (M_1), rate of mutation 2 (M_2), rate of mutation 3 (M_3), rate of mutation 4 (M_4), number of elite chromosomes that are guaranteed to pass from one generation to the

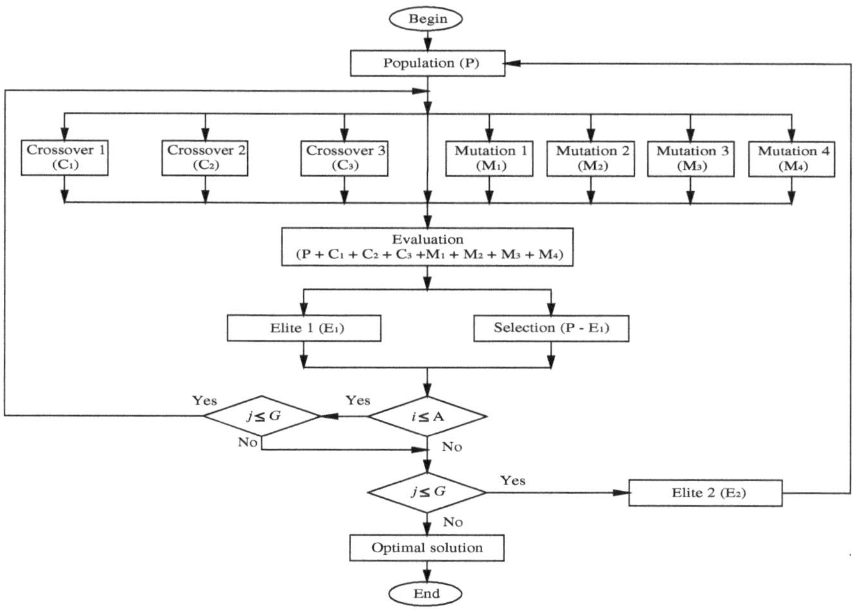

Fig. 4.4 Structure of the proposed GA

next (E_1), number of elite chromosomes that are reserved and then put into the next initial population when the search process of the proposed GA is restarted (E_2), and adaptive restart condition in term of number of successive generations within which quality of the best chromosome obtained so far is not improved (A). It should be noted that the parameters P, C_1, C_2, C_3, M_1, M_2, M_3, M_4, E_1, E_2 and A are shown in Fig. 4.4. In addition, three notations *i*, *j* and *G* in Fig. 4.4 represent (1) the current number of successive generations within which the quality of the best chromosome obtained so far is not improved, (2) the current number of generations of the proposed GA and (3) the maximum number of generations of the proposed GA, respectively. Yang and El-Haik (2003) can be consulted for details of Taguchi experimental design method.

For the sake of implementation simplicity, the pseudo code of the proposed GA in Matlab is shown in Fig. 4.5 where texts in bold are the functions in Matlab, and the rest of the pseudo code are the user-built functions, and the meanings of P, C_1, C_2, C_3, M_1, M_2, M_3, M_4, E_1, E_2, A, G, *i* and *j* were clearly explained above. Although, it is not possible to show all details, the pseudo code can provide some main ideas on how to implement the proposed GA. To fully understand the code, a certain background in Matlab programming is required. In addition, as shown in Fig. 4.4, the three crossover and four mutation operations of the proposed GA can be done either in parallel or sequential. Herein they are done in sequential as shown in Fig. 4.5.

The developed Taguchi experimental design based tuning approach consists of four steps, namely *generating Taguchi experimental design*, *conducting the*

```
INPUT: P,C1,C2,C3,M1,M2,M3,M4,E1,E2,A,G
--------------------------------------------------------------------------------
Elite2(1:E2) = Chromosome(E2)
j = 1
while j ≤ G
|       Population(1:E2) = Elite2(E2)
|       Population(E2+1:P) = Chromosome(P-E2)
|       i = 1
|       while i ≤ A
|       |   Population(P+1:P+C1) = Crossover1(C1,P)
|       |   Population(P+C1+1:P+C1+C2) = Crossover2(C2,P)
|       |   Population(P+C1+C2+1:P+C1+C2+C3) = Crossover3(C3,P)
|       |   Population(P+C1+C2+C3+1:P+C1+C2+C3+M1) = Mutation1(M1,P)
|       |   Population(P+C1+C2+C3+M1+1:P+C1+C2+C3+M1+M2) = Mutation2(M2,P)
|       |   Population(P+C1+C2+C3+M1+M2+1:P+C1+C2+C3+M1+M2+M3) = Mutation3(M3,P)
|       |   Population(P+C1+C2+C3+M1+M2+M3+1:P+C1+C2+C3+M1+M2+M3+M4) = Mutation4(M4,P)
|       |   Fitness(1:P+C1+C2+C3+M1+M2+M3+M4) = Evaluation(P+C1+C2+C3+M1+M2+M3+M4)
|       |   Elite1(1:E1) = Top(E1,P+C1+C2+C3+M1+M2+M3+M4)
|       |   Population(1:E1) = Elite1(E1)
|       |   Population(E1+1:P) = Selection(P-E1,P+C1+C2+C3+M1+M2+M3+M4-E1)
|       |   i = i + 1
|       |   j = j + 1
|       |   if j > G
|       |   |   break
|       |   end
|       end
|       Elite2(1:E2) = Top(E2,P)
end
Optimal_Solution = Best(1,P)
--------------------------------------------------------------------------------
OUTPUT: Optimal_Solution
```

Fig. 4.5 The pseudo code of the proposed GA in Matlab

experiments, *analysing the experimental data* and *selecting the parameter values*. Details of the steps are described as follows.

Step 1: Generating Taguchi experimental design

As a large number of parameters are involved, Taguchi experimental design (Yang and El-Haik 2003) is the best tool to employ herein. In Taguchi experimental design, each parameter has a number of so called levels; for example, population size may have three levels: 50, 100 and 150. In principle, the more levels the better; but it is not always practically possible to increase the number of experimental parameter levels, because a much larger number of experiments will be required. With 11 parameters P, C_1, C_2, C_3, M_1, M_2, M_3, M_4, E_1, E_2 and A, according to Minitab, Taguchi Orthogonal Array Design has only two types of design, namely two-level and three-level design where in the former each parameter has two levels and in the latter three levels. Therefore, to maximise the number of experimental parameter levels, Taguchi Orthogonal Array Design L27 (3^{11}), a three-level design, is chosen whose details are shown in Tables 4.22 and 4.23.

To cover the most promising ranges identified by (1) several preliminary tests and (2) nearly ten years of experience working with GA of the author, three experimental levels of the parameters are chosen as shown in Table 4.22. It is noted that crossover and mutation rates herein are in terms of the number of chromosomes to be crossed and mutated in one generation, respectively. In addition, adaptive restart condition is the number of successive generations within which quality of the best chromosome obtained so far is not improved.

There are 27 experiments in Taguchi Orthogonal Array Design L27 (3^{11}), and in each experiment, the proposed GA has a different set of the parameters as shown in Table 4.23; for example, the proposed GA in the first experiment has the following parameter set: P = 50, $C_1 = 10$, $C_2 = 10$, $C_3 = 10$, $M_1 = 2$, $M_2 = 2$, $M_3 = 2$, $M_4 = 2$, $E_1 = 1$, $E_2 = 1$ and A = 10. How these experiments should be conducted

Table 4.22 Parameters of the proposed GA and their experimental levels

No.	Parameter	Code	Level		
			1	2	3
1	Population size	P	50	100	150
2	Rate of crossover 1	C1	10	40	70
3	Rate of crossover 2	C2	10	40	70
4	Rate of crossover 3	C3	10	40	70
5	Rate of mutation 1	M1	2	6	10
6	Rate of mutation 2	M2	2	6	10
7	Rate of mutation 3	M3	2	6	10
8	Rate of mutation 4	M4	2	6	10
9	No. of elite chromosomes (type 1)	E1	1	5	9
10	No. of elite chromosomes (type 2)	E2	1	5	9
11	Adaptive restart condition	A	10	30	50

Table 4.23 Experimental design L27 (3^{11}) generated by Minitab

Experiment	Parameter of the proposed GA										
	P	C1	C2	C3	M1	M2	M3	M4	E1	E2	A
1	50	10	10	10	2	2	2	2	1	1	10
2	50	10	10	10	6	6	6	6	5	5	30
3	50	10	10	10	10	10	10	10	9	9	50
4	50	40	40	40	2	2	2	6	5	5	50
5	50	40	40	40	6	6	6	10	9	9	10
6	50	40	40	40	10	10	10	2	1	1	30
7	50	70	70	70	2	2	2	10	9	9	30
8	50	70	70	70	6	6	6	2	1	1	50
9	50	70	70	70	10	10	10	6	5	5	10
10	100	10	40	70	2	6	10	2	5	9	10
11	100	10	40	70	6	10	2	6	9	1	30
12	100	10	40	70	10	2	6	10	1	5	50
13	100	40	70	10	2	6	10	6	9	1	50
14	100	40	70	10	6	10	2	10	1	5	10
15	100	40	70	10	10	2	6	2	5	9	30
16	100	70	10	40	2	6	10	10	1	5	30
17	100	70	10	40	6	10	2	2	5	9	50
18	100	70	10	40	10	2	6	6	9	1	10
19	150	10	70	40	2	10	6	2	9	5	10
20	150	10	70	40	6	2	10	6	1	9	30
21	150	10	70	40	10	6	2	10	5	1	50
22	150	40	10	70	2	10	6	6	1	9	50
23	150	40	10	70	6	2	10	10	5	1	10
24	150	40	10	70	10	6	2	2	9	5	30
25	150	70	40	10	2	10	6	10	5	1	30
26	150	70	40	10	6	2	10	2	9	5	50
27	150	70	40	10	10	6	2	6	1	9	10

and what kind of data needed to be collected to assist the users with selecting the best parameter set will be presented in the next Step.

Step 2: Conducting the experiments

The proposed GA with 27 different parameter sets as shown in Table 4.23 are independently run; and the objective function values, i.e. the total costs of all product orders under consideration, obtained in these 27 experiments are collected for further processing in the next Step. Each experiment should be repeated for a number of times, say five, to increase the consistency of the experiment response.

To make a fair comparison between different parameter sets, termination criterion of the proposed GA, i.e. the computing time, must be set exactly the same in every experiment.

The proposed VCIM production scheduling model is a stochastic model. However, for the purpose of tuning the parameters of the proposed GA, a deterministic version of the proposed VCIM production scheduling model is used, in which all stochastic parameters are forced to take their nominal values, i.e. deterministic model. There are two reasons for using the deterministic model herein, instead of the stochastic one. First, from the proposed GA point of view, the problems in the stochastic and deterministic models are almost the same; the only difference is in the evaluation part presented in Sect. 4.5 where the objective function value in the deterministic model can be directly calculated while in the stochastic model it must be estimated using Monte Carlo simulation. Second, using the deterministic version of the proposed model to tune the parameters of the proposed GA can significantly reduce the computing time, because it does not involve Monte Carlo simulation.

Step 3: Analysing the experimental data

To determine the effects of the parameters on the proposed GA performance, ANOVA analysis is carried out where, according to Yang and El-Haik (2003), the relative importance of an effect to the experiment response is presented by the corresponding *F value*, the larger the more important. In addition, *p value* is used to determine whether an effect is statistically significant to the experiment response. An effect is commonly considered significant if its *p value* is less than 0.05. In this thesis, ANOVA analysis is done by Minitab software. The parameter effects, found in ANOVA analysis, will serve as a foundation for selecting the best parameter values of the proposed GA. The proposed procedure for selecting those values will be presented in the next Section.

Step 4: Selecting the parameter values

Based on the result of ANOVA analysis in Step 3, the parameters of the proposed GA are classified into two groups: *insignificant* and *significant*. The values of insignificant parameters are selected based on the main-effect chart generated by Minitab, in which the experimental parameter levels associated with the highest fitness values should be chosen. For significant parameters, based on the main-effect chart generated by Minitab, the experimental parameter levels associated with the highest fitness values are first chosen as the starting points; and then, for each individual parameter, Hill Climbing technique with the corresponding chosen starting point is applied to search for the optimal parameter value. Hill Climbing is a local optimisation technique, which starts with a given initial solution and then attempts to improve it by incrementally altering its characteristics (Taborda and Zdravkovic 2012).

4.8 Termination Criterion

Termination criteria in the proposed GA can be (1) maximum number of generations, (2) maximum number of objective function evaluations, and (3) maximum computing time. Generally speaking, the three termination criteria are equivalent because they all stop the GA with a pre-set value. Nevertheless, the author prefers, and used, the computing time because it gives the users a clear idea on when to get the results.

4.9 Concluding Remarks

In this chapter, an innovative GA with unique chromosome representation, modified genetic operations as well as novel algorithm structure was developed to solve a class of VCIM production scheduling problems. This class of problems, i.e. dynamic multi-objective optimisation problems with uncertainties, exist in the most comprehensive VCIM production scheduling model to date, which was introduced in Chap. 3.

To encode the solution, a special chromosome representation with four different parts representing four different sub-problems, namely component supplier selection, assembly agent selection, collaborative component shipment scheduling and collaborative product shipment scheduling, was developed. To better explore the solution space, three different types of crossover operations and four different types of mutation operations were introduced into the proposed GA. To enhance the global search capability, a novel GA structure with an adaptive stop-and-restart-with-memory mechanism was developed herein. Finally, a systematic approach based on Taguchi experimental design was proposed to tune the proposed GA by selecting the best combination of its parameters to maximise the performance.

References

Aghaei, J., Amjady, N., & Shayanfar, H. A. (2011). Multi-objective electricity market clearing considering dynamic security by lexicographic optimization and augmented epsilon constraint method. *Applied Soft Computing, 11*(4), 3846–3858.

Boudissa, E., & Bounekhla, M. (2012). Genetic algorithm with dynamic selection based on quadratic ranking applied to induction machine parameters estimation. *Electric Power Components and Systems, 40*(10), 1089–1104.

Brandimarte, P. (2014). *Handbook in Monte Carlo simulation: Applications in financial engineering, risk management and economics*. Hoboken, New Jersey: Wiley.

Cabrera, G. G, Ehrgott, M., Mason, A., & Philpott, A. (2014). Multi-objective optimisation of positively homogeneous functions and an application in radiation therapy. *Operations Research Letters, 42*(4), 268–272.

Castelli, M., & Vanneschi, L. (2014). Genetic algorithm with variable neighborhood search for the optimal allocation of goods in shop shelves. *Operations Research Letters, 42*(5), 355–360.
Chen, A., & Xu, X. (2012). Goal programming approach to solving network design problem with multiple objectives and demand uncertainty. *Expert Systems with Applications, 39*(4), 4160–4170.
Chiandussi, G., Codegone, M., Ferrero, S., & Varesio, F. E. (2012). Comparison of multi-objective optimization methodologies for engineering applications. *Computers & Mathematics with Applications, 63*(5), 912–942.
Copado-Méndez, P. J., Pozo, C., Guillén-Gosálbez, G., & Jiménez, L. (2016). Enhancing the ϵ-constraint method through the use of objective reduction and random sequences: Application to environmental problems. *Computers & Chemical Engineering, 87,* 36–48.
Gen, M., & Cheng, R. (1997). *Genetic algorithms and engineering design.* New York: Wiley.
Giagkiozis, I., & Fleming, P. J. (2015). Methods for multi-objective optimization: An analysis. *Information Sciences, 293,* 338–350.
Kang, L., Liu, Y., & Liang, X. (2015). Multi-objective optimization of heat exchanger networks based on analysis of minimum temperature difference and accumulated CO_2 emissions. *Applied Thermal Engineering, 87,* 736–748.
Khalili-Damghani, K., & Amiri, M. (2012). Solving binary-state multi-objective reliability redundancy allocation series-parallel problem using efficient epsilon-constraint, multi-start partial bound enumeration algorithm, and DEA. *Reliability Engineering & System Safety, 103,* 35–44.
Kitayama, S., & Yamazaki, K. (2012). Compromise point incorporating trade-off ratio in multi-objective optimization. *Applied Soft Computing, 12*(8), 1959–1964.
Laumanns, M., Thiele, L., & Zitzler, E. (2006). An efficient, adaptive parameter variation scheme for metaheuristics based on the epsilon-constraint method. *European Journal of Operational Research, 169*(3), 932–942.
Li, Y., He, H., Wang, Y., Xu, X., & Jiao, L. (2015). An improved multiobjective estimation of distribution algorithm for environmental economic dispatch of hydrothermal power systems. *Applied Soft Computing, 28,* 559–568.
Maity, S., Roy, A., & Maiti, M. (2015). A modified genetic algorithm for solving uncertain constrained solid travelling salesman problems. *Computers & Industrial Engineering, 83,* 273–296.
Marler, R. T., & Arora, J. S. (2009). The weighted sum method for multi-objective optimization: New insights. *Structural and Multidisciplinary Optimization, 41*(6), 853–862.
Mavrotas, G. (2009). Effective implementation of the ε-constraint method in multi-objective mathematical programming problems. *Applied Mathematics and Computation, 213*(2), 455–465.
Rastegar, N., & Khorram, E. (2014). A combined scalarizing method for multiobjective programming problems. *European Journal of Operational Research, 236*(1), 229–237.
Ren, Z. W., San, Y., & Chen, J. F. (2007). Hybrid simplex-improved genetic algorithm for global numerical optimization. *Acta Automatica Sinica, 33*(1), 91–95.
Rezvani, A., Gandomkar, M., Izadbakhsh, M., & Ahmadi, A. (2015). Environmental/economic scheduling of a micro-grid with renewable energy resources. *Journal of Cleaner Production, 87,* 216–226.
Shopova, E. G., & Vaklieva-Bancheva, N. G. (2006). BASIC—A genetic algorithm for engineering problems solution. *Computers & Chemical Engineering, 30*(8), 1293–1309.
Taborda, D. M. G., & Zdravkovic, L. (2012). Application of a Hill-Climbing technique to the formulation of a new cyclic nonlinear elastic constitutive model. *Computers and Geotechnics, 43,* 80–91.
Yang, K., & El-Haik, B. (2003). *Design for Six Sigma: A roadmap for product development.* New York: McGraw-Hill.
Zhao, J., & Wang, L. (2011). Center based genetic algorithm and its application to the stiffness equivalence of the aircraft wing. *Expert Systems with Applications, 38*(5), 6254–6261.

Chapter 5
Case Study

5.1 Introduction

As mentioned before, currently, there are two major limitations in the VCIM production scheduling: lack of a comprehensive scheduling model and a robust optimisation solution method. To overcome the limitations, an innovative production scheduling model for VCIM systems and a modified Genetic Algorithm (GA) for solving the optimisation problem in that model were presented in Chaps. 3 and 4, respectively. To test the effectiveness of the proposed VCIM production scheduling model and to verify the robustness of the proposed GA, a comprehensive case study is carried out in this Chapter where a VCIM system producing a variety of electric bikes is considered. With a dataset collected from the current market, performance of the proposed approach is compared to those of the traditional approaches. A number of insights obtained from the case study are also discussed herein.

5.2 Problem Description

Electric bikes are suitable products that can be used to test the performances of the proposed VCIM production scheduling model as well as the proposed GA, because of three following reasons. First, electric bikes have become increasingly popular throughout the world, due to technology, economics, and government policy (Weinert et al. 2007). Second, small/medium size enterprises anywhere in the world could be assembly agents of the VCIM system, since it is relatively easy to set up an assembly agent to assemble electric bikes. Third, components of electric bikes can be produced and supplied by many manufacturing agents, locally and/or globally distributed.

S. D. Dao, *Modelling and Intelligent Optimisation of Production Scheduling in VCIM Systems*, Springer Theses, https://doi.org/10.1007/978-3-319-72113-2_5

Suppose a VCIM system providing six different types of electric bikes, namely mountain electric bike, city electric bike, folding electric bike, three wheel electric bike, sport electric bike, and snow electric bike, is established. Components of each bike are shown in Table 5.1. The VCIM system is assumed to have 10 component suppliers and seven assembly agents, and each day receives a number of product orders from its customers. Component suppliers, assembly agents as well as the customers of the VCIM system are locally and/or globally distributed. More details about the VCIM system, the products as well as the customers, will be presented in Sect. 5.3. After a certain period of time, say one day, the VCIM system needs to do the production scheduling in order to fulfil the received customer orders. Without loss of generality, let us assume that the VCIM system is currently required to do the production scheduling for six customer orders, each with one type of electric bike, as shown in Table 5.2.

The scheduling questions here are: (1) what is the allocation of the components of the requested bikes to the component suppliers; (2) what is the allocation of the requested bikes to the assembly agents; (3) what is the shipping schedule to transport the components and bikes to the required destinations; so that a temporary VCIM production system can be formed to simultaneously fulfil all of the six customer orders with minimum total cost, minimum average completion time, and maximum average reliability.

5.3 Data Collection

Costs, average release times, sizes and weights of the bike components as well as sizes and weights of the finished bikes were estimated based on the data available at Alibaba.com on 16 April 2016, as shown in Tables 5.3, 5.4, 5.5, 5.6, 5.7, 5.8, 5.9, 5.10, 5.11, 5.12, 5.13, 5.14, 5.15, 5.16, 5.17. Alibaba.com is the e-commerce giant, running B2B (business to business), B2C (business to customer) and C2C (customer to customer) e-marketplace that connects small and medium-size buyers and suppliers all over the world (Qing and Xue 2009). It should be noted that the data at Alibaba.com is changing from day to day; and since the VCIM system does not exist yet, it is not legal to display actual names of component suppliers as well as assembly agents in this case study. In addition, notation "–" in every Table in this thesis means "not available". For example, in Table 5.3, the cost of mountain electric bike battery produced by component supplier CS1 is not available; in other words, component supplier CS1 does not supply the battery.

Assembly cost and time of the electric bikes are shown in Tables 5.18 and 5.19, which were estimated based on the author's survey conducted in several electric bike factories in Thai Nguyen city, Vietnam. Locations of the component suppliers, assembly agents and customers, which are assumed to be in large cities distributed worldwide, are expressed in terms of latitudes and longitudes as shown in Tables 5.20, 5.21, 5.22. Shipping times and shipping costs were estimated based on guidelines of Australia Post (2015) and a research of Lair (2008) as shown in

Table 5.1 Components of six electric bikes

No.	Product	Product component												
		Motor	Battery	Charger	Frame	Wheels	Breaks	Saddle	Chain	Crankset	Pedals	Display	Lights	Others
1	Mountain electric bike	+	+	+	+	+	+	+	+	+	+	+	–	+
2	City electric bike	+	+	+	+	+	+	+	+	+	+	+	+	+
3	Folding electric bike	+	+	+	+	+	+	+	+	+	+	+	+	+
4	Three wheel electric bike	+	+	+	+	+	+	+	+	+	+	–	–	+
5	Sport electric bike	+	+	+	+	+	+	+	+	+	+	+	+	+
6	Snow electric bike	+	+	+	+	+	+	+	+	+	+	+	+	+

Note "+" means "available", and "–" means "not available"

Table 5.2 Customer orders to be fulfilled

No.	Customer order	Product	Quantity
1	1	Mountain electric bike	1
2	2	City electric bike	1
3	3	Folding electric bike	1
4	4	Three wheel electric bike	1
5	5	Sport electric bike	1
6	6	Snow electric bike	1

Tables 5.23, 5.24, 5.25, 5.26, 5.27. Finally, qualities of the bike components as well as qualities of the assembly operations for assembling the bikes, in terms of defects per million opportunities (DPMO), are assumed in Tables 5.28, 5.29, 5.30, 5.31, 5.32, 5.33, 5.34. It should be noted that in terms of Sigma Level, these qualities are about 5.1–5.5 Sigma, which are very good quality targets in today's manufacturing industry, according to (Castanheira et al. 2011; Saghaei et al. 2012).

5.4 Implementation of the Proposed Approach

The proposed approach, including the developed VCIM production scheduling model described in Chap. 3 and the developed GA described in Chap. 4, is applied to solve this case study problem. As mentioned in Sect. 4.5, in the proposed approach, ε-constraint method is used to deal with multiple objective problem, where the total cost objective function is to be minimised while two other objective functions, i.e. minimum average completion time and maximum average reliability, are transformed into constraints with pre-set threshold values. Without loss of generality, it is assumed that the decision makers set a standard performance of the VCIM system, with maximum average completion time of 14.0 days, maximum standard deviation of the completion times of 3.5 days and minimum average reliability of 0.9987 (DPMO $\leq$ 1300 or sigma level $\geq$ 4.5).

In this case study, to evaluate the effectiveness of the proposed approach, the proposed VCIM production scheduling model is compared to two other scheduling models and the proposed GA is compared to three other optimisation algorithms. In order to make these comparisons, parameters of the proposed GA are needed and will be presented in the next Section.

5.4.1 *Parameters of the Proposed Genetic Algorithm*

As mentioned in Sect. 4.7, parameters of the proposed GA are systematically tuned by Taguchi experimental design based approach. The experimental data are shown

Table 5.3 Estimated costs of the mountain electric bike components ($)

No.	C. Supplier	Product component												
		Motor	Battery	Charger	Frame	Wheels	Breaks	Saddle	Chain	Crankset	Pedals	Display	Lights	Others
1	CS1	113.8	–	–	103.8	38.8	18.6	4.7	–	17.9	2.5	7.2	–	59.6
2	CS2	134.0	150.8	5.1	188.5	42.9	–	1.0	1.6	–	–	6.3	–	39.4
3	CS3	–	192.3	8.2	–	30.5	17.3	2.8	1.5	–	2.9	14.5	–	–
4	CS4	108.1	215.6	4.3	179.6	59.5	8.9	–	1.7	16.7	2.2	8.7	–	68.8
5	CS5	105.9	222.3	7.8	109.9	–	13.9	–	–	14.4	1.0	–	–	64.7
6	CS6	124.9	–	–	126.2	33.2	5.3	4.1	2.0	18.3	–	9.0	–	33.4
7	CS7	–	160.9	5.6	133.5	41.2	11.4	2.3	1.2	–	2.7	10.2	–	–
8	CS8	117.0	213.2	7.4	–	35.9	–	4.1	1.1	11.7	–	11.8	–	44.8
9	CS9	129.3	–	–	113.7	44.7	7.4	2.9	–	18.6	2.7	–	–	–
10	CS10	111.2	163.4	3.6	172.1	–	7.7	–	1.1	19.9	1.4	13.6	–	53.9

Table 5.4 Estimated costs of the city electric bike components ($)

No.	C. Supplier	Product component												
		Motor	Battery	Charger	Frame	Wheels	Breaks	Saddle	Chain	Crankset	Pedals	Display	Lights	Others
1	CS1	137.6	159.9	9.5	110.7	–	11.3	1.7	1.4	15.1	2.1	–	7.5	61.6
2	CS2	112.8	164.2	8.4	165.4	57.6	–	–	–	18.8	2.3	13.7	5.4	44.7
3	CS3	125.3	166.8	6.4	149.4	31.6	14.0	2.9	–	–	1.1	9.1	–	–
4	CS4	135.0	–	–	177.9	52.1	12.1	–	1.8	11.5	2.2	10.0	8.1	33.5
5	CS5	–	181.7	6.1	–	38.1	15.4	2.4	1.6	12.0	1.7	6.5	8.6	60.9
6	CS6	–	–	–	190.4	42.7	–	3.4	–	14.1	–	–	7.1	38.2
7	CS7	127.4	171.8	6.6	–	46.4	–	–	1.9	17.5	2.0	12.0	–	–
8	CS8	106.9	–	–	133.4	58.3	5.5	4.0	2.0	18.3	1.4	–	8.3	52.1
9	CS9	107.5	239.3	8.7	169.9	–	6.0	2.0	1.2	–	1.2	14.1	6.4	39.2
10	CS10	112.9	–	–	119.8	–	9.8	4.7	1.1	13.2	–	7.1	–	–

Table 5.5 Estimated costs of the folding electric bike components ($)

No.	C. Supplier	Product component												
		Motor	Battery	Charger	Frame	Wheels	Breaks	Saddle	Chain	Crankset	Pedals	Display	Lights	Others
1	CS1	–	–	–	103.1	39.0	13.0	2.1	1.7	15.3	1.3	–	8.0	49.4
2	CS2	112.7	168.4	9.5	174.4	–	–	4.1	1.1	10.9	1.4	10.9	5.0	–
3	CS3	140.7	171.2	6.0	–	–	11.1	1.8	1.5	–	–	12.4	5.2	61.3
4	CS4	–	157.7	4.3	148.0	46.2	17.3	2.2	–	11.4	2.3	15.0	–	–
5	CS5	146.5	241.4	9.3	190.5	–	15.8	1.4	1.9	16.8	1.6	8.6	7.4	41.8
6	CS6	117.5	–	–	161.0	50.0	19.5	–	–	15.0	–	9.7	–	39.5
7	CS7	109.8	205.8	6.1	–	35.3	–	3.7	1.4	–	2.4	–	7.9	51.2
8	CS8	–	181.3	3.8	185.9	33.8	9.9	–	–	15.0	2.0	12.9	7.8	33.7
9	CS9	130.8	166.6	4.8	180.5	60.0	6.6	2.7	1.7	–	–	13.4	–	46.2
10	CS10	123.7	–	–	157.7	35.1	14.2	–	1.5	10.5	1.9	6.9	6.2	–

Table 5.6 Estimated costs of the three wheel electric bike components ($)

No.	C. Supplier	Product component												
		Motor	Battery	Charger	Frame	Wheels	Breaks	Saddle	Chain	Crankset	Pedals	Display	Lights	Others
1	CS1	117.6	248.8	7.2	–	31.0	16.7	3.6	1.3	–	1.2	–	–	–
2	CS2	–	–	–	124.0	46.8	11.4	–	–	15.6	2.0	–	–	61.4
3	CS3	129.3	175.8	7.2	188.7	56.5	6.4	–	1.6	19.3	2.7	–	–	41.7
4	CS4	127.5	189.7	8.0	102.9	50.1	9.0	4.8	1.3	17.0	–	–	–	54.1
5	CS5	145.9	157.4	4.6	149.0	35.7	–	1.8	1.0	15.8	1.5	–	–	68.6
6	CS6	114.3	218.4	3.8	–	–	9.2	3.8	–	–	1.4	–	–	47.3
7	CS7	–	190.2	5.1	197.9	43.8	11.6	1.9	–	18.8	–	–	–	57.8
8	CS8	137.7	248.3	5.2	171.3	59.4	12.9	1.5	1.4	19.9	2.3	–	–	60.3
9	CS9	–	–	–	150.0	–	–	–	1.7	10.0	1.8	–	–	–
10	CS10	128.4	212.1	6.6	–	55.7	18.1	2.8	1.4	18.7	1.4	–	–	56.2

Table 5.7 Estimated costs of the sport electric bike components ($)

No.	C. Supplier	Product component												
		Motor	Battery	Charger	Frame	Wheels	Breaks	Saddle	Chain	Crankset	Pedals	Display	Lights	Others
1	CS1	103.8	165.4	3.6	106.0	49.3	12.8	2.8	1.7	16.1	2.9	14.2	7.2	34.4
2	CS2	–	–	–	168.2	41.3	–	3.6	1.4	19.9	1.2	6.9	6.9	67.4
3	CS3	126.5	166.1	8.6	104.2	35.7	14.6	4.1	–	–	–	12.7	–	37.5
4	CS4	139.0	225.8	3.2	–	42.8	19.4	–	–	14.8	1.3	12.6	8.2	40.6
5	CS5	146.7	237.1	9.5	152.2	44.5	–	3.6	1.4	18.0	–	–	7.9	–
6	CS6	106.5	185.1	8.1	109.7	33.6	15.1	2.7	1.0	12.3	2.2	7.7	–	49.5
7	CS7	128.4	–	–	181.8	47.7	9.3	4.4	1.3	15.0	2.1	11.4	5.3	60.8
8	CS8	–	179.4	7.0	–	36.8	–	–	–	–	1.1	8.7	5.4	–
9	CS9	100.6	203.1	4.7	172.2	41.5	15.4	2.0	1.3	15.7	2.9	7.2	–	40.9
10	CS10	116.9	233.2	6.2	115.0	–	6.0	3.5	1.2	18.5	2.5	–	8.8	31.5

Table 5.8 Estimated costs of the snow electric bike components ($)

No.	C. Supplier	Product component												
		Motor	Battery	Charger	Frame	Wheels	Breaks	Saddle	Chain	Crankset	Pedals	Display	Lights	Others
1	CS1	108.1	209.7	9.7	166.0	37.6	8.8	–	1.8	17.4	2.5	14.1	7.7	56.9
2	CS2	–	183.5	6.8	–	38.7	8.4	3.2	–	15.9	–	6.6	–	–
3	CS3	115.6	179.9	6.6	197.3	–	15.0	4.5	1.9	–	2.7	8.2	7.9	48.1
4	CS4	126.4	195.3	4.6	164.9	38.0	17.7	2.1	1.4	16.7	2.9	–	5.4	54.4
5	CS5	108.3	–	–	180.0	54.7	–	2.3	1.8	10.8	3.0	10.0	–	32.4
6	CS6	130.1	–	–	145.4	59.5	–	1.5	–	16.3	–	6.1	7.6	42.6
7	CS7	113.1	205.8	7.8	–	–	15.1	4.8	1.8	16.6	2.6	14.1	6.3	60.9
8	CS8	–	224.3	5.8	–	40.3	5.1	3.6	–	17.3	–	7.8	7.6	57.9
9	CS9	134.5	192.4	5.6	108.3	47.5	14.0	–	1.4	18.9	1.4	–	–	35.0
10	CS10	137.4	–	–	113.3	33.2	10.8	–	1.2	19.8	1.8	8.8	7.3	–

Table 5.9 Average release times of the mountain electric bike components (day)

No.	C.S.	Motor		Battery		Charger		Frame		Wheels		Breaks		Saddle		Chain		Crankset		Pedals		Display		Lights		Others	
		μ	σ	μ	σ	μ	σ	μ	σ	μ	σ	μ	σ	μ	σ	μ	σ	μ	σ	μ	σ	μ	σ	μ	σ	μ	σ
1	CS1	1.6	0.1	–	–	–	–	4.7	0.4	3.9	0.3	2.3	0.2	2.9	0.2	–	–	2.8	0.2	2.5	0.2	2.2	0.2	–	–	3.2	0.3
2	CS2	4.9	0.4	2.3	0.2	2.8	0.2	4.1	0.3	2.6	0.2	–	–	2.1	0.2	2.9	0.2	–	–	–	–	3.0	0.2	–	–	4.4	0.4
3	CS3	–	–	5.1	0.4	1.4	0.1	–	–	2.6	0.2	2.9	0.2	1.6	0.1	2.3	0.2	–	–	2.4	0.2	2.1	0.2	–	–	–	–
4	CS4	2.9	0.2	2.2	0.2	1.5	0.1	2.7	0.2	1.7	0.1	2.3	0.2	–	–	2.9	0.2	2.4	0.2	2.4	0.2	2.0	0.2	–	–	2.8	0.2
5	CS5	4.2	0.3	5.6	0.5	1.3	0.1	2.8	0.2	–	–	2.6	0.2	–	–	–	–	3.6	0.3	1.9	0.2	–	–	–	–	1.2	0.1
6	CS6	1.6	0.1	–	–	–	–	2.2	0.2	2.9	0.2	1.9	0.2	1.4	0.1	2.4	0.2	2.7	0.2	–	–	1.5	0.1	–	–	1.7	0.1
7	CS7	–	–	2.0	0.2	2.7	0.2	3.0	0.2	3.0	0.2	1.9	0.1	2.8	0.2	1.6	0.1	–	–	1.7	0.1	2.0	0.2	–	–	–	–
8	CS8	4.7	0.4	2.3	0.2	2.2	0.2	–	–	2.2	0.2	–	–	1.7	0.1	2.3	0.2	4.3	0.3	–	–	2.8	0.2	–	–	2.3	0.2
9	CS9	4.2	0.3	–	–	–	–	4.3	0.3	2.1	0.2	1.2	0.1	2.2	0.2	–	–	2.0	0.2	1.5	0.1	–	–	–	–	–	–
10	CS10	4.8	0.4	3.4	0.3	1.3	0.1	4.2	0.3	–	–	1.3	0.1	–	–	1.1	0.1	3.5	0.3	1.4	0.1	2.7	0.2	–	–	1.5	0.1

Note μ = mean, σ = standard deviation

Table 5.10 Average release times of the city electric bike components (day)

No.	C.S.	Motor		Battery		Charger		Frame		Wheels		Breaks		Saddle		Chain		Crankset		Pedals		Display		Lights		Others	
		μ	σ	μ	σ	μ	σ	μ	σ	μ	σ	μ	σ	μ	σ	μ	σ	μ	σ	μ	σ	μ	σ	μ	σ	μ	σ
1	CS1	3.6	0.3	2.8	0.2	2.7	0.2	3.6	0.3	–	–	1.3	0.1	1.8	0.1	1.5	0.1	3.3	0.3	2.6	0.2	–	–	1.4	0.1	5.0	0.4
2	CS2	1.1	0.1	5.2	0.4	2.2	0.2	2.5	0.2	3.7	0.3	–	–	–	–	–	–	3.2	0.3	1.9	0.1	2.2	0.2	2.5	0.2	3.2	0.3
3	CS3	4.4	0.4	3.9	0.3	1.7	0.1	4.2	0.3	3.7	0.3	2.7	0.2	2.9	0.2	–	–	–	–	2.8	0.2	1.5	0.1	–	–	–	–
4	CS4	4.7	0.4	–	–	–	–	3.1	0.3	3.4	0.3	2.6	0.2	–	–	2.7	0.2	2.1	0.2	1.8	0.1	2.3	0.2	3.6	0.3	5.0	0.4
5	CS5	–	–	5.6	0.4	1.8	0.1	–	–	1.3	0.1	1.1	0.1	3.5	0.3	1.7	0.1	2.3	0.2	2.5	0.2	1.2	0.1	1.8	0.1	2.2	0.2
6	CS6	–	–	–	–	–	–	4.8	0.4	1.8	0.1	–	–	3.9	0.3	–	–	1.5	0.1	–	–	–	–	1.6	0.1	2.7	0.2
7	CS7	4.0	0.3	4.8	0.4	1.5	0.1	–	–	2.0	0.2	–	–	–	–	2.4	0.2	4.8	0.4	2.6	0.2	2.3	0.2	–	–	–	–
8	CS8	2.6	0.2	–	–	–	–	3.2	0.3	3.0	0.2	1.8	0.1	2.0	0.2	1.0	0.1	3.6	0.3	2.5	0.2	–	–	2.9	0.2	4.1	0.3
9	CS9	3.6	0.3	2.9	0.2	1.4	0.1	3.5	0.3	–	–	2.3	0.2	2.8	0.2	2.2	0.2	–	–	1.8	0.1	2.8	0.2	2.3	0.2	4.3	0.3
10	CS10	1.7	0.1	–	–	–	–	3.3	0.3	–	–	2.3	0.2	1.3	0.1	1.8	0.1	3.6	0.3	–	–	3.0	0.2	–	–	–	–

Note μ = mean, σ = standard deviation

Table 5.11 Average release times of the folding electric bike components (day)

No.	C.S.	Motor		Battery		Charger		Frame		Wheels		Breaks		Saddle		Chain		Crankset		Pedals		Display		Lights		Others	
		μ	σ	μ	σ	μ	σ	μ	σ	μ	σ	μ	σ	μ	σ	μ	σ	μ	σ	μ	σ	μ	σ	μ	σ	μ	σ
1	CS1	–	–	–	–	–	–	1.8	0.1	1.3	0.1	1.6	0.1	3.7	0.3	2.8	0.2	3.2	0.3	2.6	0.2	–	–	3.8	0.3	1.7	0.1
2	CS2	1.1	0.1	1.3	0.1	1.1	0.1	2.2	0.2	–	–	–	–	3.6	0.3	1.0	0.1	3.6	0.3	2.9	0.2	2.2	0.2	1.2	0.1	–	–
3	CS3	2.1	0.2	3.7	0.3	2.8	0.2	–	–	–	–	1.0	0.1	3.5	0.3	1.9	0.2	–	–	–	–	2.9	0.2	1.3	0.1	1.2	0.1
4	CS4	–	–	4.9	0.4	2.9	0.2	1.9	0.2	3.3	0.3	3.0	0.2	1.8	0.1	–	–	3.9	0.3	2.3	0.2	2.2	0.2	–	–	–	–
5	CS5	1.4	0.1	5.7	0.5	2.0	0.2	4.4	0.4	–	–	1.3	0.1	2.8	0.2	1.9	0.2	3.1	0.2	1.9	0.2	1.0	0.1	1.5	0.1	2.3	0.2
6	CS6	4.3	0.3	–	–	–	–	1.8	0.1	3.7	0.3	1.2	0.1	–	–	–	–	5.0	0.4	–	–	1.2	0.1	–	–	1.7	0.1
7	CS7	3.8	0.3	3.8	0.3	1.7	0.1	–	–	3.7	0.3	–	–	2.3	0.2	1.6	0.1	–	–	2.5	0.2	–	–	2.7	0.2	1.8	0.1
8	CS8	–	–	3.3	0.3	2.8	0.2	1.7	0.1	2.0	0.2	1.4	0.1	–	–	–	–	1.4	0.1	1.3	0.1	2.0	0.2	1.2	0.1	4.6	0.4
9	CS9	4.8	0.4	1.1	0.1	1.7	0.1	1.9	0.2	3.1	0.2	2.0	0.2	1.5	0.1	1.9	0.2	–	–	–	–	2.7	0.2	–	–	3.7	0.3
10	CS10	1.1	0.1	–	–	–	–	2.7	0.2	1.6	0.1	1.7	0.1	–	–	1.1	0.1	1.3	0.1	3.0	0.2	1.4	0.1	3.2	0.3	–	–

Note μ = mean, σ = standard deviation

Table 5.12 Average release times of the three wheel electric bike components (day)

No.	C.S.	Motor		Battery		Charger		Frame		Wheels		Breaks		Saddle		Chain		Crankset		Pedals		Display		Lights		Others	
		μ	σ	μ	σ	μ	σ	μ	σ	μ	σ	μ	σ	μ	σ	μ	σ	μ	σ	μ	σ	μ	σ	μ	σ	μ	σ
1	CS1	2.8	0.2	1.8	0.1	2.6	0.2	–	–	1.1	0.1	2.9	0.2	2.3	0.2	1.4	0.1	–	–	2.0	0.2	–	–	–	–	–	–
2	CS2	–	–	–	–	–	–	4.7	0.4	3.2	0.3	2.8	0.2	–	–	–	–	2.8	0.2	2.8	0.2	–	–	–	–	1.4	0.1
3	CS3	4.1	0.3	2.6	0.2	1.5	0.1	2.7	0.2	2.5	0.2	1.1	0.1	–	–	1.9	0.2	2.5	0.2	2.2	0.2	–	–	–	–	4.0	0.3
4	CS4	4.2	0.3	3.6	0.3	1.8	0.1	1.7	0.1	2.4	0.2	2.5	0.2	2.4	0.2	1.3	0.1	4.1	0.3	–	–	–	–	–	–	3.9	0.3
5	CS5	1.7	0.1	1.8	0.1	1.2	0.1	4.6	0.4	3.7	0.3	–	–	3.1	0.2	1.7	0.1	3.5	0.3	1.4	0.1	–	–	–	–	3.2	0.3
6	CS6	3.0	0.2	4.0	0.3	1.3	0.1	–	–	–	–	1.8	0.1	3.1	0.2	–	–	–	–	1.8	0.1	–	–	–	–	1.7	0.1
7	CS7	–	–	2.3	0.2	2.9	0.2	2.8	0.2	2.9	0.2	2.1	0.2	2.9	0.2	–	–	4.7	0.4	–	–	–	–	–	–	3.4	0.3
8	CS8	3.6	0.3	4.3	0.3	2.9	0.2	1.4	0.1	3.6	0.3	2.9	0.2	1.1	0.1	2.5	0.2	4.9	0.4	2.7	0.2	–	–	–	–	2.2	0.2
9	CS9	–	–	–	–	–	–	2.0	0.2	–	–	–	–	–	–	1.5	0.1	1.8	0.1	2.6	0.2	–	–	–	–	–	–
10	CS10	4.0	0.3	4.7	0.4	1.1	0.1	–	–	2.7	0.2	3.0	0.2	2.0	0.2	2.8	0.2	1.6	0.1	1.6	0.1	–	–	–	–	1.9	0.1

Note μ = mean, σ = standard deviation

Table 5.13 Average release times of the sport electric bike components (day)

No.	C.S.	Motor		Battery		Charger		Frame		Wheels		Breaks		Saddle		Chain		Crankset		Pedals		Display		Lights		Others	
		μ	σ	μ	σ	μ	σ	μ	σ	μ	σ	μ	σ	μ	σ	μ	σ	μ	σ	μ	σ	μ	σ	μ	σ	μ	σ
1	CS1	2.1	0.2	3.3	0.3	1.5	0.1	3.4	0.3	1.5	0.1	1.6	0.1	2.6	0.2	1.5	0.1	3.8	0.3	2.1	0.2	2.1	0.2	3.2	0.3	4.6	0.4
2	CS2	–	–	–	–	–	–	2.0	0.2	1.7	0.1	–	–	3.0	0.2	2.5	0.2	1.4	0.1	1.2	0.1	2.3	0.2	1.2	0.1	1.3	0.1
3	CS3	3.6	0.3	2.1	0.2	2.6	0.2	3.4	0.3	3.7	0.3	2.3	0.2	2.2	0.2	–	–	–	–	–	–	1.1	0.1	–	–	2.0	0.2
4	CS4	1.7	0.1	5.6	0.4	1.0	0.1	–	–	1.1	0.1	2.1	0.2	–	–	–	–	3.1	0.2	1.3	0.1	2.2	0.2	3.8	0.3	1.2	0.1
5	CS5	1.5	0.1	1.8	0.1	1.1	0.1	1.9	0.2	2.5	0.2	–	–	3.2	0.3	1.2	0.1	4.4	0.4	–	–	–	–	4.0	0.3	–	–
6	CS6	3.0	0.2	5.1	0.4	1.3	0.1	1.5	0.1	1.5	0.1	2.3	0.2	3.9	0.3	2.2	0.2	2.9	0.2	2.0	0.2	1.1	0.1	–	–	1.1	0.1
7	CS7	4.8	0.4	–	–	–	–	2.2	0.2	3.9	0.3	1.4	0.1	2.6	0.2	2.4	0.2	2.6	0.2	1.4	0.1	2.0	0.2	3.4	0.3	4.6	0.4
8	CS8	–	–	6.0	0.5	2.5	0.2	–	–	3.1	0.3	–	–	–	–	–	–	–	–	2.0	0.2	1.4	0.1	2.5	0.2	–	–
9	CS9	3.3	0.3	1.4	0.1	2.3	0.2	2.7	0.2	2.5	0.2	3.0	0.2	1.3	0.1	1.9	0.1	4.0	0.3	1.3	0.1	1.2	0.1	–	–	1.4	0.1
10	CS10	1.9	0.2	3.2	0.3	1.9	0.2	3.0	0.2	–	–	1.3	0.1	2.8	0.2	2.3	0.2	3.1	0.2	1.1	0.1	–	–	2.2	0.2	2.2	0.2

Note μ = mean, σ = standard deviation

Table 5.14 Average release times of the snow electric bike components (day)

No.	C.S.	Motor		Battery		Charger		Frame		Wheels		Breaks		Saddle		Chain		Crankset		Pedals		Display		Lights		Others	
		μ	σ	μ	σ	μ	σ	μ	σ	μ	σ	μ	σ	μ	σ	μ	σ	μ	σ	μ	σ	μ	σ	μ	σ	μ	σ
1	CS1	4.0	0.3	1.5	0.1	2.1	0.2	1.3	0.1	1.2	0.1	1.1	0.1	–	–	2.3	0.2	2.4	0.2	2.7	0.2	1.3	0.1	1.4	0.1	2.8	0.2
2	CS2	–	–	5.8	0.5	1.6	0.1	–	–	3.0	0.2	2.1	0.2	2.3	0.2	–	–	1.6	0.1	–	–	1.4	0.1	–	–	–	–
3	CS3	3.0	0.2	1.0	0.1	2.5	0.2	4.2	0.3	–	–	2.8	0.2	1.3	0.1	2.3	0.2	–	–	2.9	0.2	1.1	0.1	3.8	0.3	5.0	0.4
4	CS4	3.8	0.3	4.9	0.4	1.4	0.1	1.1	0.1	1.2	0.1	2.3	0.2	1.8	0.1	2.9	0.2	2.0	0.2	2.4	0.2	–	–	1.9	0.2	2.3	0.2
5	CS5	4.6	0.4	–	–	–	–	4.7	0.4	2.6	0.2	–	–	1.5	0.1	1.4	0.1	1.2	0.1	2.2	0.2	1.6	0.1	–	–	2.2	0.2
6	CS6	4.8	0.4	–	–	–	–	3.9	0.3	1.3	0.1	–	–	1.8	0.1	–	–	4.0	0.3	–	–	2.1	0.2	2.0	0.2	1.2	0.1
7	CS7	3.2	0.3	1.4	0.1	1.7	0.1	–	–	–	–	1.9	0.2	2.3	0.2	1.5	0.1	2.0	0.2	2.8	0.2	2.4	0.2	2.4	0.2	2.2	0.2
8	CS8	–	–	3.0	0.2	2.3	0.2	–	–	3.5	0.3	3.0	0.2	2.6	0.2	–	–	2.8	0.2	–	–	2.0	0.2	2.9	0.2	1.2	0.1
9	CS9	1.6	0.1	2.3	0.2	2.6	0.2	1.9	0.2	3.2	0.3	1.3	0.1	–	–	2.2	0.2	3.8	0.3	1.0	0.1	–	–	–	–	3.0	0.2
10	CS10	2.0	0.2	–	–	–	–	2.8	0.2	1.4	0.1	2.7	0.2	–	–	1.9	0.2	2.4	0.2	2.7	0.2	1.9	0.2	3.5	0.3	–	–

Note μ = mean, σ = standard deviation

Table 5.15 Estimated sizes of the electric bike components (mm)

No.	Product	Product component						
		Motor	Battery	Charger	Frame	Wheels	Breaks	Saddle
1	Mountain electric bike	185 * 185 * 210	90 * 120 * 420	135 * 61 * 35	1020 * 150 * 550	610 * 610 * 500	400 * 300 * 50	265 * 170 * 100
2	City electric bike	185 * 185 * 210	90 * 120 * 420	140 * 65 * 40	1050 * 150 * 570	610 * 610 * 500	400 * 400 * 50	282 * 136 * 110
3	Folding electric bike	185 * 185 * 210	80 * 110 * 400	135 * 61 * 35	1020 * 150 * 550	610 * 610 * 500	400 * 300 * 40	270 * 150 * 110
4	Three wheel electric bike	195 * 195 * 220	95 * 130 * 430	170 * 90 * 50	1120 * 500 * 550	610 * 610 * 700	400 * 400 * 50	280 * 150 * 120
5	Sport electric bike	190 * 190 * 210	90 * 120 * 420	135 * 61 * 40	1100 * 140 * 550	610 * 610 * 500	400 * 300 * 50	282 * 136 * 110
6	Snow electric bike	185 * 185 * 210	95 * 130 * 430	135 * 62 * 40	1100 * 150 * 550	610 * 610 * 500	400 * 400 * 50	270 * 150 * 110

No.	Product	Product component						
		…	Chain	Crankset	Pedals	Display	Lights	Others
1	Mountain electric bike	…	200 * 60 * 30	270 * 180 * 80	115 * 110 * 50	120 * 90 * 60	–	700 * 500 * 120
2	City electric bike	…	200 * 60 * 30	270 * 180 * 90	105 * 100 * 50	130 * 90 * 60	460 * 380 * 400	600 * 450 * 100
3	Folding electric bike	…	200 * 60 * 30	270 * 180 * 70	115 * 110 * 50	120 * 80 * 60	450 * 370 * 380	650 * 500 * 110
4	Three wheel electric bike	…	200 * 70 * 40	280 * 180 * 90	120 * 110 * 60	–	–	700 * 500 * 120
5	Sport electric bike	…	200 * 65 * 30	270 * 180 * 80	115 * 110 * 50	130 * 90 * 50	455 * 380 * 410	600 * 450 * 110
6	Snow electric bike	…	200 * 60 * 30	270 * 180 * 80	125 * 120 * 60	120 * 85 * 60	465 * 370 * 390	550 * 500 * 120

Table 5.16 Estimated weights of the electric bike components (kg)

No.	Product	Product component												
		Motor	Battery	Charger	Frame	Wheels	Breaks	Saddle	Chain	Crankset	Pedals	Display	Lights	Others
1	Mountain electric bike	5.1	4.1	0.9	4.3	5.1	0.9	0.5	0.5	1.4	0.7	0.4	–	3.5
2	City electric bike	4.4	5.3	0.6	3.0	5.0	0.5	0.4	0.6	1.0	0.6	0.4	0.2	4.9
3	Folding electric bike	4.4	5.2	0.8	3.8	5.8	0.7	0.4	0.5	1.1	0.6	0.3	0.2	4.7
4	Three wheel electric bike	5.2	5.1	0.8	4.2	5.6	0.7	0.3	0.6	1.0	0.7	–	–	3.5
5	Sport electric bike	5.4	5.5	0.7	2.3	5.5	0.8	0.5	0.5	1.2	0.6	0.5	0.2	3.8
6	Snow electric bike	4.9	5.4	0.5	2.4	4.1	0.8	0.4	0.4	1.0	0.6	0.4	0.3	4.2

Table 5.17 Estimated sizes and weights of the finished bikes

No.	Product	Size (mm)	Weight (kg)
1	Mountain electric bike	1590 * 800 * 300	29.5
2	City electric bike	1760 * 760 * 260	35.0
3	Folding electric bike	1480 * 740 * 260	30.5
4	Three wheel electric bike	1500 * 680 * 580	33.2
5	Sport electric bike	1450 * 800 * 270	33.0
6	Snow electric bike	1610 * 800 * 290	31.0

Table 5.18 Estimated assembly costs of the electric bikes ($)

No.	Product	Assembly agent						
		AA1	AA2	AA3	AA4	AA5	AA6	AA7
1	Mountain electric bike	37.5	50.8	–	52.4	–	50.4	58.0
2	City electric bike	30.2	49.7	37.7	64.5	86.3	42.5	–
3	Folding electric bike	–	–	36.0	53.8	79.4	49.3	50.9
4	Three wheel electric bike	32.0	55.4	43.8	–	–	36.4	45.1
5	Sport electric bike	47.6	–	47.6	76.5	93.7	48.5	63.9
6	Snow electric bike	26.8	67.1	48.5	55.9	95.2	–	54.8

in Table 5.35. For details about the experimental layout as well as the notations, e.g. P, C_1, C_2, etc., it is advised to refer to Sect. 4.7. In addition, the deterministic version of the proposed VCIM production scheduling model is used to tune the parameters of the proposed GA. For the reasons why the deterministic version is used to tune the parameters, it is advised to refer to Sect. 4.7.

As can be seen from Table 5.35, the computing time of each experiment was set to be 15 min and each experiment was repeated 5 times. ANOVA analysis and main-effect chart, generated by Minitab, are shown in Table 5.36 and Fig. 5.1. The *p values* in Table 5.36 indicate that with 95% confidence level, parameters P, C_1, C_2, M_1, M_3, E_1, and E_2 have significant effects on the performance of the proposed GA. As mentioned in Sect. 4.7, Hill Climbing technique can be used to further tune these significant parameters. However, for the sake of simplicity, the further tuning was not done herein. Based on the main-effect chart in Fig. 5.1, parameters of the proposed GA were selected and shown in Table 5.37.

5.4.2 Monte Carlo Simulation

As mentioned in Chaps. 3 and 4, the developed VCIM production scheduling model is a stochastic model and Monte Carlo simulation is used to deal with the uncertainties. One of the issues in Monte Carlo simulation is sample size which

Table 5.19 Estimated assembly times of the electric bikes (h)

No.	Product	AA1		AA2		AA3		AA4		AA5		AA6		AA7	
		μ	σ	μ	σ	μ	σ	μ	σ	μ	σ	μ	σ	μ	σ
1	Mountain electric bike	2.2	0.2	2.8	0.3	–	–	1.9	0.2	–	–	2.0	0.2	3.2	0.3
2	City electric bike	2.7	0.3	2.9	0.3	3.1	0.3	3.5	0.3	2.3	0.2	2.1	0.2	–	–
3	Folding electric bike	–	–	–	–	1.8	0.2	1.6	0.2	2.7	0.3	1.9	0.2	3.3	0.3
4	Three wheel electric bike	1.8	0.2	3.2	0.3	2.2	0.2	–	–	–	–	1.5	0.2	3.4	0.3
5	Sport electric bike	3.3	0.3	–	–	3.4	0.3	2.3	0.2	1.9	0.2	2.4	0.2	2.7	0.3
6	Snow electric bike	2.8	0.3	1.8	0.2	3.1	0.3	2.3	0.2	2.2	0.2	–	–	1.7	0.2

Note μ = mean, σ = standard deviation

Table 5.20 Locations of the component suppliers

	Component supplier									
	CS1	CS2	CS3	CS4	CS5	CS6	CS7	CS8	CS9	CS10
Latitude	39.92	13.80	13.83	30.00	35.68	−15.87	−27.50	−33.80	52.52	41.87
Longitude	116.38	77.30	100.48	31.28	139.73	−47.92	153.17	18.47	13.33	−87.64

Table 5.21 Locations of the assembly agents

	Assembly agent						
	AA1	AA2	AA3	AA4	AA5	AA6	AA7
Latitude	31.23	1.37	35.75	41.30	32.79	−33.43	25.03
Longitude	121.45	103.75	51.50	2.09	−96.79	−70.67	121.63

Table 5.22 Locations of the customers

	Customer					
	CT1	CT2	CT3	CT4	CT5	CT6
Latitude	39.92	−34.77	33.23	51.50	40.75	23.13
Longitude	116.38	138.54	44.37	−0.12	−73.99	−82.38

herein is estimated by a function named *Sample-Size-for-Estimation* in Minitab, from which with estimated standard deviation of the average completion time of 3.1 days (based on preliminary tests), margin of error for confidence interval of 1.2 days, confidence level of 95%, and confidence interval of "two-sided", the sample size of Monte Carlo simulation was estimated to be 29.

Monte Carlo simulation is very time-consuming, especially for the large-scale problems like this case study. Traditionally, Monte Carlo simulation is included in a genetic algorithm, from the beginning of search process; which means Monte Carlo simulation is involved in all generations of the algorithm. This inclusion strategy (named strategy 1) is not very effective for solving large-scale problems, because it requires a massive computing time. Running the proposed GA, with Monte Carlo simulation included from the first generation, and for 400 generations would take 25.2 h.

To overcome this obstacle, a new inclusion strategy (named strategy 2) is proposed herein, which is to include Monte Carlo simulation in only a certain number of generations at the end of the search process of the proposed GA. More specifically, in early generations, the proposed GA works along with the deterministic version of the proposed VCIM production scheduling model, in which Monte Carlo simulation is not involved; in the late generations, the proposed GA works along with the stochastic version of the model where Monte Carlo simulation is involved. It should be noted that the whole population in the last generation of the deterministic stage will be transferred to be the population in the first generation of the

Table 5.23 Average shipping times between the component suppliers (day)

C.S.	CS1		CS2		CS3		CS4		CS5		CS6		CS7		CS8		CS9		CS10	
	μ	σ	μ	σ	μ	σ	μ	σ	μ	σ	μ	σ	μ	σ	μ	σ	μ	σ	μ	σ
CS1	0.0	0.0	1.0	0.3	0.7	0.2	1.6	0.5	0.4	0.1	3.5	1.1	1.8	0.5	2.7	0.8	1.5	0.5	2.2	0.7
CS2	1.0	0.3	0.0	0.0	0.5	0.2	1.0	0.3	1.4	0.4	2.9	0.9	2.0	0.6	1.7	0.5	1.5	0.4	2.8	0.9
CS3	0.7	0.2	0.5	0.2	0.0	0.0	1.5	0.5	1.0	0.3	3.5	1.0	1.5	0.5	2.1	0.6	1.8	0.5	2.9	0.9
CS4	1.6	0.5	1.0	0.3	1.5	0.5	0.0	0.0	2.0	0.6	2.1	0.6	3.0	0.9	1.5	0.5	0.6	0.2	2.1	0.6
CS5	0.4	0.1	1.4	0.4	1.0	0.3	2.0	0.6	0.0	0.0	3.7	1.1	1.5	0.4	3.1	0.9	1.9	0.6	2.1	0.6
CS6	3.5	1.1	2.9	0.9	3.5	1.0	2.1	0.6	3.7	1.1	0.0	0.0	3.1	0.9	1.4	0.4	2.0	0.6	1.6	0.5
CS7	1.8	0.5	2.0	0.6	1.5	0.5	3.0	0.9	1.5	0.4	3.1	0.9	0.0	0.0	2.4	0.7	3.3	1.0	3.0	0.9
CS8	2.7	0.8	1.7	0.5	2.1	0.6	1.5	0.5	3.1	0.9	1.4	0.4	2.4	0.7	0.0	0.0	2.0	0.6	2.8	0.9
CS9	1.5	0.5	1.5	0.4	1.8	0.5	0.6	0.2	1.9	0.6	2.0	0.6	3.3	1.0	2.0	0.6	0.0	0.0	1.5	0.4
CS10	2.2	0.7	2.8	0.9	2.9	0.9	2.1	0.6	2.1	0.6	1.6	0.5	3.0	0.9	2.8	0.9	1.5	0.4	0.0	0.0

Note μ = mean, σ = standard deviation

Table 5.24 Average shipping times between the suppliers and assembly agents (day)

C.S./A.A.	AA1		AA2		AA3		AA4		AA5		AA6		AA7	
	μ	σ	μ	σ	μ	σ	μ	σ	μ	σ	μ	σ	μ	σ
CS1	0.3	0.1	0.9	0.3	1.2	0.3	1.8	0.6	2.3	0.7	4.0	1.2	0.4	0.1
CS2	1.0	0.3	0.7	0.2	0.7	0.2	1.6	0.5	3.1	0.9	3.4	1.0	1.0	0.3
CS3	0.6	0.2	0.3	0.1	1.1	0.3	2.0	0.6	3.0	0.9	3.7	1.1	0.5	0.2
CS4	1.7	0.5	1.7	0.5	0.4	0.1	0.6	0.2	2.3	0.7	2.7	0.8	1.8	0.5
CS5	0.4	0.1	1.1	0.3	1.6	0.5	2.2	0.7	2.2	0.6	3.6	1.1	0.4	0.1
CS6	3.8	1.1	3.4	1.0	2.5	0.7	1.7	0.5	1.6	0.5	0.6	0.2	3.9	1.2
CS7	1.5	0.5	1.3	0.4	2.7	0.8	3.5	1.1	2.8	0.8	2.5	0.7	1.4	0.4
CS8	2.7	0.8	2.0	0.6	1.8	0.5	1.8	0.5	2.9	0.9	1.7	0.5	2.6	0.8
CS9	1.8	0.5	2.1	0.6	0.7	0.2	0.3	0.1	1.7	0.5	2.6	0.8	1.9	0.6
CS10	2.4	0.7	3.1	0.9	2.2	0.7	1.5	0.4	0.3	0.1	1.8	0.5	2.5	0.7

Note μ = mean, σ = standard deviation

Table 5.25 Average shipping times between the assembly agents and customers (day)

A.A./ C.T.	CT1		CT2		CT3		CT4		CT5		CT6	
	μ	σ	μ	σ	μ	σ	μ	σ	μ	σ	μ	σ
AA1	0.3	0.1	1.6	0.5	1.5	0.4	1.9	0.6	2.5	0.7	2.8	0.8
AA2	0.9	0.3	1.1	0.3	1.5	0.4	2.3	0.7	3.2	1.0	3.6	1.1
AA3	1.2	0.3	2.5	0.7	0.1	0.0	0.9	0.3	2.1	0.6	2.5	0.7
AA4	1.8	0.6	3.4	1.0	0.8	0.2	0.2	0.1	1.3	0.4	1.6	0.5
AA5	2.3	0.7	3.1	0.9	2.4	0.7	1.6	0.5	0.5	0.1	0.4	0.1
AA6	4.0	1.2	2.5	0.7	2.9	0.9	2.4	0.7	1.7	0.5	1.3	0.4
AA7	0.4	0.1	1.4	0.4	1.5	0.5	2.0	0.6	2.6	0.8	2.9	0.9

Note μ = mean, σ = standard deviation

Table 5.26 Average shipping times between the customers (day)

C.T.	CT1		CT2		CT3		CT4		CT5		CT6	
	μ	σ	μ	σ	μ	σ	μ	σ	μ	σ	μ	σ
CT1	0.0	0.0	1.8	0.5	1.3	0.4	1.7	0.5	2.3	0.7	2.7	0.8
CT2	1.8	0.5	0.0	0.0	2.6	0.8	3.4	1.0	3.6	1.1	3.3	1.0
CT3	1.3	0.4	2.6	0.8	0.0	0.0	0.9	0.3	2.0	0.6	2.4	0.7
CT4	1.7	0.5	3.4	1.0	0.9	0.3	0.0	0.0	1.2	0.3	1.6	0.5
CT5	2.3	0.7	3.6	1.1	2.0	0.6	1.2	0.3	0.0	0.0	0.4	0.1
CT6	2.7	0.8	3.3	1.0	2.4	0.7	1.6	0.5	0.4	0.1	0.0	0.0

Note μ = mean, σ = standard deviation

Table 5.27 Fixed and variable shipping cost coefficients

Fixed shipping cost coefficient ($/km)	Variable shipping cost coefficient ($/gram)
0.01	0.002

stochastic stage and in this transfer, if a chromosome is not feasible, it will be replaced by a feasible one randomly generated.

Experimental results in Table 5.38 demonstrate the effectiveness of the proposed inclusion strategy (strategy 2). In these experiments, the two strategies are compared to each other, when solving this case study problem, with the total computing time of 180 min. The proposed GA with each inclusion strategy was independently run for 5 times. In strategy 2, the proposed GA worked along with the deterministic version of the scheduling model for 90 min, worked along with the stochastic version of the model for the rest of the computing time. As can be seen from Table 5.38, strategy 2 can provide much better solutions, on average 36.5% better, compared to strategy 1. It should be noted that in the rest of this thesis strategy 2 will be used, 90 min along with the deterministic and 90 min along with the stochastic model.

5.4.3 *Performance Evaluation*

There are two performance evaluations here, i.e. performance of the proposed VCIM production scheduling model against the traditional scheduling models and performance of the proposed GA against the traditional optimisation algorithms. Details of the two performance evaluations will be presented in the next two Sections.

5.4.3.1 Performance of the Proposed VCIM Production Scheduling Model

Performances of three VCIM production scheduling models are compared to each other as shown in Table 5.39. Model 3 is the proposed VCIM production scheduling model as described in Chap. 3, in which the collaborative shipment scheduling as well as multiple product orders are simultaneously considered. Model 2 is exactly the same as Model 3, except that the product orders are processed separately. Model 1 is exactly the same as Model 3, except for the followings: (1) the collaborative shipment scheduling is not considered and (2) the product orders are processed separately. The three scheduling models and the proposed GA were applied to solve this case study problem. When changing the scheduling models, only "evaluation" component of the proposed GA will be changed, and the rest will remain the same because Models 1 & 2 are special cases of Model 3.

Table 5.28 Assumed qualities of the mountain electric bike components (DPMO)

No.	C. Supplier	Product component												
		Motor	Battery	Charger	Frame	Wheels	Breaks	Saddle	Chain	Crankset	Pedals	Display	Lights	Others
1	CS1	40	–	–	110	124	84	89	–	50	78	136	–	38
2	CS2	38	64	146	65	112	–	53	33	–	–	138	–	39
3	CS3	–	97	125	–	55	129	47	80	–	91	138	–	–
4	CS4	37	126	91	84	84	45	–	122	80	85	137	–	106
5	CS5	37	78	78	64	–	44	–	–	89	47	–	–	138
6	CS6	57	–	–	149	114	48	116	89	124	–	79	–	81
7	CS7	–	119	38	86	81	88	142	117	–	105	146	–	–
8	CS8	132	129	38	–	94	–	103	119	92	–	121	–	119
9	CS9	85	–	–	93	149	149	126	–	68	130	–	–	–
10	CS10	98	108	99	136	–	46	–	38	134	90	50	–	128

Table 5.29 Assumed qualities of the city electric bike components (DPMO)

No.	C. Supplier	Product component												
		Motor	Battery	Charger	Frame	Wheels	Breaks	Saddle	Chain	Crankset	Pedals	Display	Lights	Others
1	CS1	63	140	68	53	–	122	86	85	147	121	–	87	32
2	CS2	123	59	143	146	133	–	–	–	118	63	103	61	44
3	CS3	136	113	114	46	113	50	40	–	–	61	75	–	–
4	CS4	117	–	–	92	132	117	–	82	62	37	69	86	98
5	CS5	–	131	97	–	62	67	125	89	52	60	46	96	63
6	CS6	–	–	–	41	119	–	148	–	99	–	–	49	124
7	CS7	147	94	83	–	140	–	–	42	128	79	81	–	–
8	CS8	78	–	–	88	132	49	55	99	49	100	–	50	142
9	CS9	65	144	93	131	–	76	86	105	–	75	93	36	119
10	CS10	145	–	–	127	–	131	52	85	134	–	57	–	–

Table 5.30 Assumed qualities of the folding electric bike components (DPMO)

No.	C. Supplier	Product component												
		Motor	Battery	Charger	Frame	Wheels	Breaks	Saddle	Chain	Crankset	Pedals	Display	Lights	Others
1	CS1	–	–	–	126	88	51	62	38	96	145	–	88	49
2	CS2	141	119	105	62	–	–	60	109	147	118	87	150	–
3	CS3	57	31	107	–	–	140	35	60	–	–	115	132	73
4	CS4	–	43	125	79	85	50	90	–	75	77	86	–	–
5	CS5	71	110	114	78	–	109	94	55	137	62	49	121	144
6	CS6	70	–	–	118	148	111	–	–	129	–	65	–	92
7	CS7	43	74	48	–	33	–	47	53	–	103	–	148	40
8	CS8	–	86	69	35	32	70	–	–	102	42	82	86	112
9	CS9	67	141	136	148	39	135	68	83	–	–	77	–	104
10	CS10	129	–	–	131	84	148	–	45	147	79	107	38	–

Table 5.31 Assumed qualities of the three wheel electric bike components (DPMO)

No.	C. Supplier	Product component												
		Motor	Battery	Charger	Frame	Wheels	Breaks	Saddle	Chain	Crankset	Pedals	Display	Lights	Others
1	CS1	37	125	138	–	133	43	125	149	–	144	–	–	–
2	CS2	–	–	–	37	74	111	–	–	110	127	–	–	70
3	CS3	63	86	64	149	50	88	–	54	114	133	–	–	35
4	CS4	115	133	67	103	70	119	70	91	130	–	–	–	75
5	CS5	116	113	43	75	132	–	135	77	109	79	–	–	126
6	CS6	136	135	53	–	–	135	92	–	–	39	–	–	65
7	CS7	–	93	103	35	145	98	84	–	88	–	–	–	52
8	CS8	79	83	62	127	66	70	113	147	101	54	–	–	108
9	CS9	–	–	–	86	–	–	–	45	89	129	–	–	–
10	CS10	121	135	111	–	134	86	148	141	72	117	–	–	38

Table 5.32 Assumed qualities of the sport electric bike components (DPMO)

No.	C. Supplier	Product component												
		Motor	Battery	Charger	Frame	Wheels	Breaks	Saddle	Chain	Crankset	Pedals	Display	Lights	Others
1	CS1	142	76	122	104	128	60	53	140	139	61	110	87	114
2	CS2	–	–	–	48	63	–	37	147	112	120	117	134	118
3	CS3	91	73	65	113	71	127	148	–	–	–	66	–	82
4	CS4	96	108	51	–	74	84	–	–	78	73	52	46	78
5	CS5	104	136	124	34	52	–	146	128	45	–	–	125	–
6	CS6	79	93	113	59	63	65	144	42	136	91	104	–	117
7	CS7	62	–	–	95	113	96	97	76	104	107	66	109	147
8	CS8	–	77	129	–	96	–	–	–	–	150	147	95	–
9	CS9	61	135	107	79	49	70	73	77	37	124	106	–	41
10	CS10	33	141	57	72	–	32	134	146	122	108	–	66	83

Table 5.33 Assumed qualities of the snow electric bike components (DPMO)

No.	C. Supplier	Product component												
		Motor	Battery	Charger	Frame	Wheels	Breaks	Saddle	Chain	Crankset	Pedals	Display	Lights	Others
1	CS1	38	51	100	31	44	110	–	111	55	107	73	39	109
2	CS2	–	89	41	–	105	106	134	–	58	–	141	–	–
3	CS3	94	105	46	100	–	129	83	93	–	70	59	77	60
4	CS4	137	48	51	59	111	35	72	96	120	132	–	98	110
5	CS5	90	–	–	106	90	–	113	138	42	81	38	–	84
6	CS6	48	–	–	47	108	–	85	–	140	–	79	117	83
7	CS7	140	51	64	–	–	50	76	103	54	131	143	128	150
8	CS8	–	82	56	–	145	128	84	–	77	–	33	62	46
9	CS9	104	122	106	110	94	101	–	149	67	138	–	–	72
10	CS10	122	–	–	85	95	110	–	115	39	142	59	148	–

Table 5.34 Assumed qualities of the assembly operations (DPMO)

No.	Product	Assembly agent						
		AA1	AA2	AA3	AA4	AA5	AA6	AA7
1	Mountain electric bike	124	108	–	55	–	81	113
2	City electric bike	92	111	77	54	65	79	–
3	Folding electric bike	–	–	31	44	52	99	131
4	Three wheel electric bike	133	32	69	–	–	65	79
5	Sport electric bike	84	–	127	32	148	104	83
6	Snow electric bike	68	134	98	113	38	–	100

As can be seen from Table 5.39, the computing time of every experiment was set exactly the same and each experiment was repeated for 5 times. It should be noted that only objective function 2 has two parameters, mean value and standard deviation, because it is affected by the stochastic time parameters, i.e. stochastic release times, stochastic assembly times and stochastic shipping times. Obviously, all three scheduling models could provide the solutions with objective functions 2 & 3 satisfying the threshold constraints. As mentioned in Sect. 5.4, the mean value and standard deviation of objective function 2 must be smaller than or equal to 14.0 and 3.5, respectively. In addition, value of objective function 3 must be greater than or equal to 0.9987.

Performances of the three scheduling models are evaluated, based on the total cost of the product orders (objective function 1), since the average completion time (objective function 2) and the average reliability (objective function 3) have been transformed into constraints with the threshold values. The experimental results in Table 5.39 reveal that the average total cost of the solutions obtained by Model 3 is \$5651.3, while that of Model 1 and Model 2 is \$6036.7 and \$5964.9, respectively. Obviously, on average, the proposed model (Model 3) could provide the solutions with 6.4 and 5.3% smaller costs, compared to Model 1 and Model 2, respectively. These cost savings confirm the effectiveness of the proposed VCIM production scheduling model against the traditional scheduling models.

5.4.3.2 Performance of the Proposed Genetic Algorithm

Three other optimisation algorithms, i.e. Cuckoo Search (Yang and Deb 2010), Particle Swarm Optimisation (Heris 2015a) and the traditional Genetic Algorithm (Heris 2015b), are used herein as benchmarks to evaluate the performance of the proposed GA. These optimisation algorithms cannot process the chromosomes of the proposed GA because those chromosomes are uniquely designed to solve the VCIM production scheduling problem. Therefore, another chromosome representation, i.e. string-like chromosome representation, is required herein. A string-like chromosome, which encodes a solution to the case study problem, and will be used

Table 5.35 Experimental data

Exp.	Parameter of the proposed GA											Computing time (min)	Total cost ($)				
	P	C1	C2	C3	M1	M2	M3	M4	E1	E2	A		Run 1	Run 2	Run 3	Run 4	Run 5
1	50	10	10	10	2	2	2	2	1	1	10	15	7534.1	7608.7	7537.5	7505.5	7672.5
2	50	10	10	10	6	6	6	6	5	5	30	15	6458.3	6476.3	6400.1	6416.9	6528.4
3	50	10	10	10	10	10	10	10	9	9	50	15	6200.8	6368.2	6341.0	6337.8	6213.6
4	50	40	40	40	2	2	2	6	5	5	50	15	6923.7	6973.5	6945.6	6740.4	6670.1
5	50	40	40	40	6	6	6	10	9	9	10	15	6403.4	6510.4	6492.3	6593.9	6732.1
6	50	40	40	40	10	10	10	2	1	1	30	15	7687.6	7639.6	7175.4	7217.9	7913.2
7	50	70	70	70	2	2	2	10	9	9	30	15	7104.3	7166.8	6945.2	6951.8	6987.3
8	50	70	70	70	6	6	6	2	1	1	50	15	7508.9	7376.1	7535.7	7672.0	7107.9
9	50	70	70	70	10	10	10	6	5	5	10	15	6787.1	6972.5	6587.3	6856.9	6743.9
10	100	10	40	70	2	6	10	2	5	9	10	15	7221.4	7424.9	7093.6	7345.0	7299.9
11	100	10	40	70	6	10	2	6	9	1	30	15	7067.3	6855.5	7430.3	7127.1	6960.9
12	100	10	40	70	10	2	6	10	1	5	50	15	7052.9	7086.3	7891.6	7657.2	7948.4
13	100	40	70	10	2	6	10	6	9	1	50	15	6921.8	7005.1	7234.9	7430.4	7520.1
14	100	40	70	10	6	10	2	10	1	5	10	15	7848.3	7750.8	8063.6	8055.0	8210.7
15	100	40	70	10	10	2	6	2	5	9	30	15	7082.3	7004.1	6847.5	6870.6	7276.2
16	100	70	10	40	2	6	10	10	1	5	30	15	7810.6	8088.6	7905.3	8082.1	7818.1
17	100	70	10	40	6	10	2	2	5	9	50	15	7771.6	7305.2	7362.4	7044.9	7429.9
18	100	70	10	40	10	2	6	6	9	1	10	15	6881.2	6924.8	7542.7	7064.2	7009.6
19	150	10	70	40	2	10	6	2	9	5	10	15	7534.8	7556.9	7396.2	7659.7	7533.0
20	150	10	70	40	6	2	10	6	1	9	30	15	8130.9	7385.3	7795.9	8386.6	8285.3
21	150	10	70	40	10	6	2	10	5	1	50	15	7663.5	7363.7	7805.6	7402.6	7402.1
22	150	40	10	70	2	10	6	6	1	9	50	15	7541.0	8117.6	8217.2	8404.0	8094.6
23	150	40	10	70	6	2	10	10	5	1	10	15	7704.2	7728.1	7520.6	7342.0	7340.8

(continued)

Table 5.35 (continued)

Exp.	Parameter of the proposed GA											Computing time (min)	Total cost ($)				
	P	C1	C2	C3	M1	M2	M3	M4	E1	E2	A		Run 1	Run 2	Run 3	Run 4	Run 5
24	150	40	10	70	10	6	2	2	9	5	30	15	7032.1	7147.3	7199.4	7435.5	7433.0
25	150	70	40	10	2	10	6	10	5	1	30	15	7410.1	7464.3	7973.4	8029.4	7393.3
26	150	70	40	10	6	2	10	2	9	5	50	15	7036.9	7102.5	7055.0	7197.2	7205.8
27	150	70	40	10	10	6	2	6	1	9	10	15	7757.0	8039.5	8615.5	8002.2	8166.2

Table 5.36 ANOVA analysis

Source	DF	Seq SS	Adj SS	Adj MS	F	p
P	2	11,296,304	11,296,304	5,648,152	113.4	0.00
C1	2	461,863	461,863	230,932	4.6	0.01
C2	2	306,192	306,192	153,096	3.1	0.05
C3	2	258,355	258,355	129,177	2.6	0.08
M1	2	1,277,964	1,277,964	638,982	12.8	0.00
M2	2	125,329	125,329	62,665	1.3	0.29
M3	2	650,918	650,918	325,459	6.5	0.00
M4	2	60,423	60,423	30,212	0.6	0.55
E1	2	15,433,902	15,433,902	7,716,951	154.9	0.00
E2	2	428,399	428,399	214,199	4.3	0.02
A	2	276,122	276,122	138,061	2.8	0.07
Error	112	5,578,854	5,578,854	49,811		
Total	134	3,615,4626				

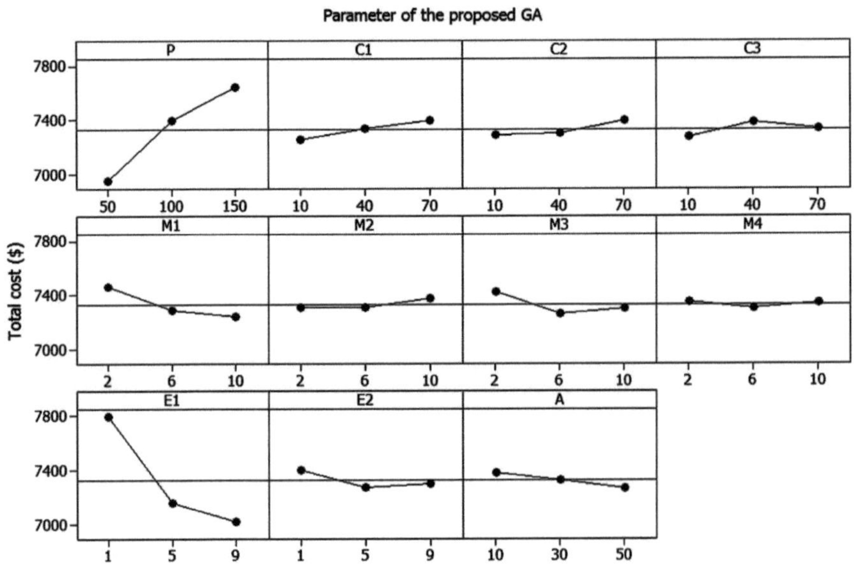

Fig. 5.1 Main-effect chart

Table 5.37 Selected parameters of the proposed GA

P	C1	C2	C3	M1	M2	M3	M4	E1	E2	A
50	10	10	10	10	2	6	6	9	5	50

Table 5.38 Strategies to include Monte Carlo simulation in the proposed GA

Run	Computing time (min)	Total cost ($)	
		Strategy 1	Strategy 2
1	180	8781.1	5626.8
2	180	8739.8	5605.2
3	180	8546.4	5603.3
4	180	9254.3	5730.7
5	180	9184.8	5690.7
Average		8901.3	5651.3

in the Cuckoo Search, Particle Swarm Optimisation and traditional Genetic Algorithm, was developed as shown in Table 5.40. For detailed explanation of the string-like chromosome, it is advised to see a supporting video, created by the author, at website: https://www.youtube.com/watch?v=TsHOpJR8VbM&feature=em-upload_owner.

Parameters of the Cuckoo Search, the Particle Swarm Optimisation and the traditional Genetic Algorithm were also tuned by the same Taguchi experimental design based approach as described in Sect. 4.7. The related experimental data, ANOVA analyses as well as main-effect charts, which support the parameter tunings of the three algorithms, are shown in Tables 5.41, 5.42, 5.43, 5.44, 5.45, 5.46, 5.47, 5.48, 5.49 and Figs. 5.2, 5.3, 5.4. It should be noted that computing time in every experiment here, i.e. 15 min, was set exactly the same as the one in the parameter tuning experiment of the proposed GA. The selected parameters of the three algorithms are shown in Tables 5.50, 5.51, 5.52.

Performances of the four optimisation algorithms, i.e. the Cuckoo Search, the Particle Swarm Optimisation, the traditional Genetic Algorithm and the proposed Genetic Algorithm, in solving the case study problem are compared to each other as shown in Table 5.53. The computing time of every algorithm was 180 min and each algorithm was independently run for 5 times. As mentioned before, only objective function 2 has two parameters, mean value and standard deviation, because it is affected by the stochastic time parameters, i.e. stochastic release times, stochastic assembly times and stochastic shipping times.

It should be noted that performances of the four optimisation algorithms are evaluated based on the total cost of the product orders (objective function 1), since the average completion time (objective function 2) and the average reliability (objective function 3) have been transformed into constraints with the threshold values. Through these threshold values, the decision makers set the standard performance of the VCIM system. In this case study, as mentioned before, mean value and standard deviation of objective function 2 must be smaller than or equal to 14.0 and 3.5, respectively. In addition, value of objective function 3 must be greater than or equal to 0.9987. All solutions obtained by the algorithms as shown in Table 5.53 have values of objective functions 2 & 3 satisfying the threshold constraints because all infeasible solutions have been removed during the evolutionary process of the solutions.

Table 5.39 Model performance comparison

Run	Comp.time (min)	Model 1			Model 2			Model 3		
		Obj1	Obj2*	Obj3	Obj1	Obj2*	Obj3	Obj1	Obj2*	Obj3
1	180	6160.4	9.8 (0.7)	0.998761	5899.9	9.2 (0.6)	0.998755	5626.8	11.3 (0.9)	0.998732
2	180	6072.7	8.0 (0.3)	0.998702	6176.7	10.1 (0.7)	0.998768	5605.2	10.9 (0.8)	0.998708
3	180	5968.4	10.6 (1.7)	0.998816	5821.7	9.6 (0.5)	0.998706	5603.3	9.6 (0.5)	0.998804
4	180	6008.7	9.3 (0.4)	0.998708	6004.4	9.9 (0.9)	0.998827	5730.7	10.7 (0.7)	0.998798
5	180	5973.5	9.4 (0.5)	0.998716	5921.8	12.5 (2.5)	0.998788	5690.7	9.9 (0.8)	0.998787
Average		6036.7	9.4 (0.7)	0.998741	5964.9	10.3 (1.0)	0.998769	5651.3	10.6 (0.8)	0.998766

Note This objective function has two parameters: mean value (standard deviation)

Table 5.40 String-like chromosome representation

Product Order 1											Product Order 2										Prod.Ship.Scheduling					
Com.Sup.Selection					AS	Com.Ship.Scheduling					Com.Sup.Selection				AS	Com.Ship.Scheduling					AA1			AA2		
Co1	Co2	Co3	Co4	Co5		3	1	5	4		Co1	Co2	Co3	Co4		2	3	1	5		Cu1	Cu2	Dir	Cu1	Cu2	Dir
3	3	1	5	4	2	4	1	2	2	5	2	3	1	5	1	1	1	3	3	5	0	1	1	1	1	0

Table 5.41 Parameters of the Cuckoo Search and their experimental levels

No.	Parameter	Code	Level		
			1	2	3
1	Number of nests	A	50	100	150
2	Discovery rate of alien eggs	B	0.25	0.50	0.75
3	Levy exponent	C	1.00	1.50	2.00

Table 5.42 Experimental layout L9 (3^3) and experimental data

Experiment	Parameter of the Cuckoo Search			Computing time (min)	Total cost ($)				
	A	B	C		Run 1	Run 2	Run 3	Run 4	Run 5
1	50	0.25	1.0	15	7570.9	7879.2	7647.7	7579.0	7614.4
2	50	0.50	1.5	15	7779.2	7957.8	7885.1	7732.3	7800.4
3	50	0.75	2.0	15	8298.2	7941.2	8172.2	8022.5	8200.6
4	100	0.25	1.5	15	7974.2	7882.3	8037.3	8032.5	7947.9
5	100	0.50	2.0	15	8314.1	8323.2	8365.6	8306.6	8247.8
6	100	0.75	1.0	15	7926.2	7960.4	7735.2	7473.8	7779.2
7	150	0.25	2.0	15	8368.6	8440.6	8437.5	8388.7	8062.5
8	150	0.50	1.0	15	7830.3	7824.2	8081.7	7961.9	7973.4
9	150	0.75	1.5	15	8131.9	8015.2	8137.9	8209.9	7969.8

Table 5.43 ANOVA analysis for tuning parameters of the Cuckoo Search

Source	DF	Seq SS	Adj SS	Adj MS	F	*p*
A	2	475,011	475,011	237,505	16.4	0.00
B	2	10,014	10,014	5007	0.4	0.71
C	2	1,691,518	1,691,518	845,759	58.4	0.00
Error	38	550,343	550,343	14,483		
Total	44	2,726,886				

The data in Table 5.53 indicate that the proposed GA outperforms the Cuckoo Search, the Particle Swarm Optimisation and the traditional Genetic Algorithm. The proposed GA found solutions with average total cost of $5651.3 (average value of Obj1), while the Cuckoo Search, the Particle Swarm Optimisation and the

Table 5.44 Parameters of the Particle Swarm Optimisation and the experimental levels

No.	Parameter	Code	Level		
			1	2	3
1	Swarm size	A	50	100	150
2	Inertia weight	B	0.5	1.0	1.5
3	Inertia weight damping ratio	C	0.5	0.7	0.9
4	Personal learning coefficient	D	1.0	1.5	2.0
5	Global learning coefficient	E	1.0	2.0	3.0
6	Velocity limit coefficient	F	0.05	0.10	0.15

traditional Genetic Algorithm found solutions with average total costs of 6846.0, 8326.6 and $5997.7, respectively. On average, solutions obtained by the proposed GA have 17.4, 32.1 and 5.8% smaller costs, compared to solutions obtained by the Cuckoo Search, the Particle Swarm Optimisation and the traditional Genetic Algorithm, respectively.

The proposed GA also found a solution with the smallest cost of $5603.3 as shown in Table 5.53. Mean value and standard deviation of completion time of the product orders in this solution are 9.6 and 0.5 days respectively. Average reliability of the products in this solution is 0.998804 which is equivalent to DPMO of 1196 or sigma level of nearly 4.6. Details of the solution with the smallest cost are shown in Tables 5.54, 5.55, 5.56. It should be noted that notations S, A and O in Table 5.55 represent the component supplier, assembly agent and product order respectively. To fully understand the solution, it is advised to refer to Sect. 4.2.

5.4.4 Sensitivity Analysis

As mentioned before, the average completion time of the product orders was set to be smaller than or equal to 14 days by the decision makers (the threshold value of objective function 2). It would be interesting to see what will happen if the maximum average completion time is shortened. To investigate this scenario, a number of experiments were carried out whose experimental data are shown in Table 5.57. As can be seen, on average, total cost of the products will increase from $5651.3 to $6418.2 if the maximum average completion time of the products decreases from 14 days to 12 days. In other words, the total cost will go up 11.9% if the maximum average completion time is reduced by two days. But if the maximum average completion time of the products goes down to 10 days, there will be no feasible solution to the problem, which means there is no solution satisfying this completion time constraint.

Table 5.45 Experimental layout L27 (3^6) and experimental data

Experiment	Parameter of the Parti. Swar. Optimi.						Computing time (min)	Total cost ($)				
	A	B	C	D	E	F		Run 1	Run 2	Run 3	Run 4	Run 5
1	50	0.5	0.5	1.0	1.0	0.05	15	8425.0	8549.7	8564.5	8362.4	8639.6
2	50	0.5	0.5	1.0	2.0	0.10	15	8553.8	8790.1	8160.5	8659.4	8413.9
3	50	0.5	0.5	1.0	3.0	0.15	15	8521.9	8622.5	8412.8	8569.0	8505.7
4	50	1.0	0.7	1.5	1.0	0.05	15	8466.3	8489.1	8478.4	8174.7	8370.5
5	50	1.0	0.7	1.5	2.0	0.10	15	8554.5	8645.1	8480.6	8341.3	8595.1
6	50	1.0	0.7	1.5	3.0	0.15	15	7956.7	8490.7	8399.6	8524.9	8644.4
7	50	1.5	0.9	2.0	1.0	0.05	15	8053.9	8500.5	8476.2	8420.9	8606.2
8	50	1.5	0.9	2.0	2.0	0.10	15	8486.7	8508.9	8498.5	8367.7	8602.5
9	50	1.5	0.9	2.0	3.0	0.15	15	8443.3	8475.6	8234.3	8505.5	8522.7
10	100	0.5	0.7	2.0	1.0	0.10	15	8421.8	8614.9	8614.0	8337.3	8565.2
11	100	0.5	0.7	2.0	2.0	0.15	15	8429.6	8279.8	8543.1	8529.1	8676.7
12	100	0.5	0.7	2.0	3.0	0.05	15	8523.7	8365.0	8152.5	8725.4	8788.4
13	100	1.0	0.9	1.0	1.0	0.10	15	8384.1	8486.6	8304.8	8482.3	8266.7
14	100	1.0	0.9	1.0	2.0	0.15	15	8512.1	8532.3	8396.7	8360.4	8586.4
15	100	1.0	0.9	1.0	3.0	0.05	15	8529.1	8294.8	8523.2	8597.0	8457.8
16	100	1.5	0.5	1.5	1.0	0.10	15	8586.4	8518.9	8043.4	8352.7	8616.8
17	100	1.5	0.5	1.5	2.0	0.15	15	8205.3	8395.9	7919.5	8570.4	8498.1
18	100	1.5	0.5	1.5	3.0	0.05	15	8805.9	8351.5	8538.7	8144.7	8263.6
19	150	0.5	0.9	1.5	1.0	0.15	15	8239.2	8645.2	8507.1	8551.9	8451.5
20	150	0.5	0.9	1.5	2.0	0.05	15	8480.8	8623.3	7960.7	8721.1	8355.6
21	150	0.5	0.9	1.5	3.0	0.10	15	8051.9	8522.9	8697.2	8209.2	8209.1
22	150	1.0	0.5	2.0	1.0	0.15	15	8447.8	8537.4	8490.7	8451.3	8450.3
23	150	1.0	0.5	2.0	2.0	0.05	15	8369.1	8399.4	8391.9	8476.4	8340.1

(continued)

Table 5.45 (continued)

Experiment	Parameter of the Parti. Swar. Optimi.						Computing time (min)	Total cost ($)				
	A	B	C	D	E	F		Run 1	Run 2	Run 3	Run 4	Run 5
24	150	1.0	0.5	2.0	3.0	0.10	15	8399.1	8211.8	8713.2	8620.3	8470.4
25	150	1.5	0.7	1.0	1.0	0.15	15	8544.3	8550.4	8493.4	8566.5	8556.0
26	150	1.5	0.7	1.0	2.0	0.05	15	8546.0	8587.2	8397.8	8590.5	8659.1
27	150	1.5	0.7	1.0	3.0	0.10	15	8164.3	8270.7	8570.4	8184.7	8661.4

Table 5.46 ANOVA analysis for tuning parameters of the Particle Swarm Optimisation

Source	DF	Seq SS	Adj SS	Adj MS	F	*p*
A	2	11,389	11,389	5694	0.2	0.83
B	2	37,021	37,021	18,510	0.6	0.55
C	2	40,066	40,066	20,033	0.7	0.52
D	2	120,489	120,489	60,244	2.0	0.14
E	2	15,409	15,409	7704	0.3	0.78
F	2	3254	3254	1627	0.1	0.95
Error	122	3,733,012	3,733,012	30,598		
Total	134	3,960,638				

Table 5.47 Parameters of the traditional Genetic Algorithm and the experimental levels

No.	Parameter	Code	Level		
			1	2	3
1	Population size	A	50	100	150
2	Crossover percentage	B	0.5	0.7	0.9
3	Mutation percentage	C	0.1	0.3	0.5
4	Extra range factor for crossover	D	0.2	0.4	0.6
5	Mutation rate	E	0.1	0.2	0.3
6	Selection method	F	Roulette wheel	Tournament	Random

Table 5.48 ANOVA analysis for tuning parameters of the traditional GA

Source	DF	Seq SS	Adj SS	Adj MS	F	*p*
A	2	16,671,936	16,671,936	8,335,968	68.2	0.00
B	2	525,720	525,720	262,860	2.2	0.12
C	2	1,688,877	1,688,877	844,438	6.9	0.00
D	2	247,100	247,100	123,550	1.0	0.37
E	2	7,873,020	7,873,020	3,936,510	32.2	0.00
F	2	5,785,577	5,785,577	2,892,789	23.7	0.00
Error	122	14,905,688	14,905,688	122,178		
Total	134	47,697,918				

Table 5.49 Experimental layout L27 (3^6) and experimental data

Experiment	Parameter of the traditional GA						Computing time (min)	Total cost ($)				
	A	B	C	D	E	F*		Run 1	Run 2	Run 3	Run 4	Run 5
1	50	0.5	0.1	0.2	0.1	1	15	6513.0	6727.2	7212.7	6976.0	6690.0
2	50	0.5	0.1	0.2	0.2	2	15	7016.8	6748.4	6844.7	7769.3	7122.4
3	50	0.5	0.1	0.2	0.3	3	15	7392.9	7430.4	8003.9	7602.4	7599.3
4	50	0.7	0.3	0.4	0.1	1	15	6391.6	6557.5	7104.6	6193.6	6581.0
5	50	0.7	0.3	0.4	0.2	2	15	7238.5	6551.6	6643.9	7139.8	7295.6
6	50	0.7	0.3	0.4	0.3	3	15	6759.5	7621.8	7440.8	6753.4	7230.4
7	50	0.9	0.5	0.6	0.1	1	15	6873.8	6924.3	6469.5	6917.1	6439.5
8	50	0.9	0.5	0.6	0.2	2	15	6528.5	7318.1	6276.0	5946.1	7180.4
9	50	0.9	0.5	0.6	0.3	3	15	7755.8	7470.0	7423.5	7657.8	8023.1
10	100	0.5	0.3	0.6	0.1	2	15	7149.7	6903.7	7038.5	6957.5	7043.3
11	100	0.5	0.3	0.6	0.2	3	15	7147.6	7288.2	7390.3	7562.1	7516.1
12	100	0.5	0.3	0.6	0.3	1	15	7549.9	8065.5	7991.8	7858.6	8261.2
13	100	0.7	0.5	0.2	0.1	2	15	6500.8	6797.6	6792.4	7211.0	6794.3
14	100	0.7	0.5	0.2	0.2	3	15	8031.9	7159.5	7327.5	7235.6	7827.4
15	100	0.7	0.5	0.2	0.3	1	15	7607.3	8115.7	8514.6	7421.2	7653.0
16	100	0.9	0.1	0.4	0.1	2	15	6749.0	6690.0	6973.0	6905.7	6604.6
17	100	0.9	0.1	0.4	0.2	3	15	8191.7	8070.6	8215.7	8354.8	7944.5
18	100	0.9	0.1	0.4	0.3	1	15	8152.0	8003.7	7904.6	8410.4	8243.3
19	150	0.5	0.5	0.4	0.1	3	15	7841.6	7273.9	7619.9	7408.6	7557.6
20	150	0.5	0.5	0.4	0.2	1	15	7604.4	6969.0	7842.7	7651.6	7761.2
21	150	0.5	0.5	0.4	0.3	2	15	7945.6	8040.5	7817.4	7471.9	8163.0
22	150	0.7	0.1	0.6	0.1	3	15	8147.6	8423.5	8237.8	7900.8	8302.5
23	150	0.7	0.1	0.6	0.2	1	15	8271.2	8344.3	8282.2	8319.3	8257.8

(continued)

Table 5.49 (continued)

Experiment	Parameter of the traditional GA						Computing time (min)	Total cost ($)				
	A	B	C	D	E	F*		Run 1	Run 2	Run 3	Run 4	Run 5
24	150	0.7	0.1	0.6	0.3	2	15	8222.9	7598.2	6894.8	7322.1	7654.2
25	150	0.9	0.3	0.2	0.1	3	15	8010.2	8265.8	7971.1	8142.7	7871.2
26	150	0.9	0.3	0.2	0.2	1	15	8097.1	7906.9	7845.6	7468.0	8104.6
27	150	0.9	0.3	0.2	0.3	2	15	8112.9	7912.5	8088.9	8086.1	7956.9

**Note* Values 1, 2, 3 in this column represent the corresponding levels of the parameter

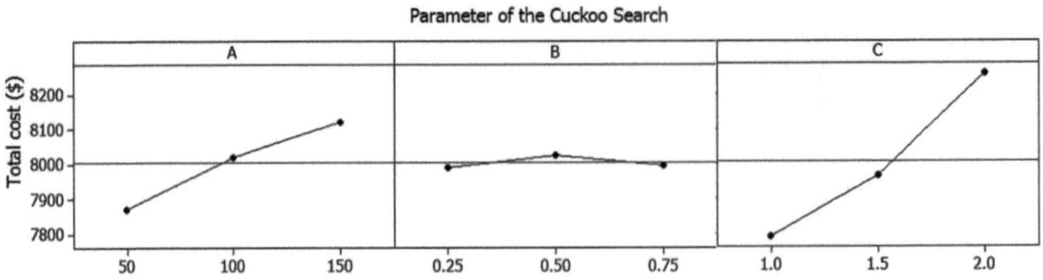

Fig. 5.2 Main-effect chart for tuning parameters of the Cuckoo Search

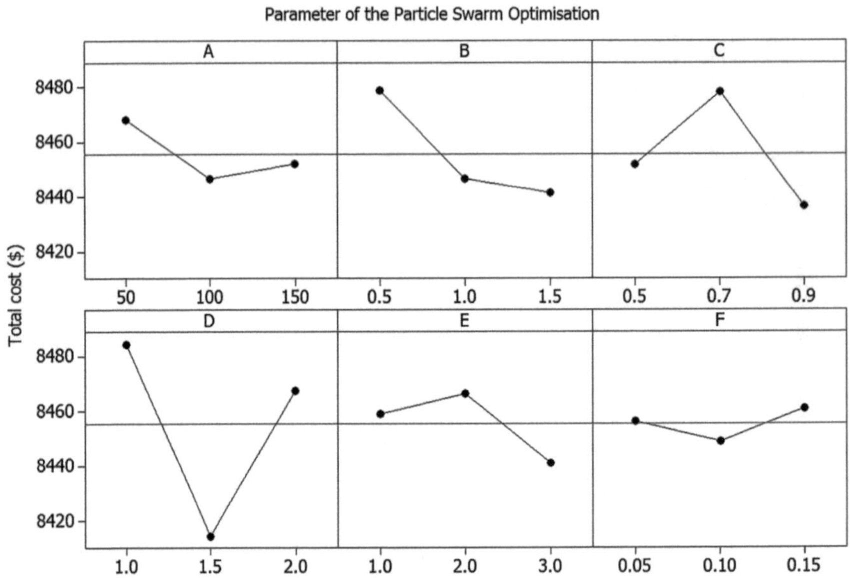

Fig. 5.3 Main-effect chart for tuning parameters of the Particle Swarm Optimisation

Again, as mentioned before, the average reliability of the products was set to be greater than or equal to 0.9987 (the threshold value of objective function 3), which means DPMO $\leq$ 1300 or sigma level $\geq$ 4.5. To investigate the effect of the reliability threshold values on the cost of the products, more experiments were conducted as shown in Table 5.58 from which it can be seen that when the reliability of the products decreases from 4.5 to 4.0 sigma level, the total cost of the products, on average, reduces from \$5651.3 to \$5558.1 (1.7% cost reduction). But if the reliability of the products increases to 5.0 sigma level, there will be no solution satisfying that reliability constraint.

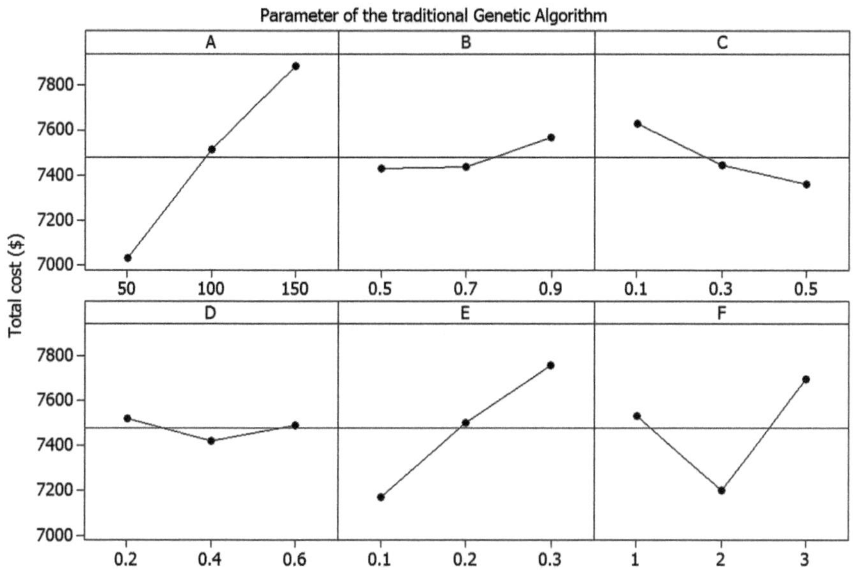

Fig. 5.4 Main-effect chart for tuning parameters of the traditional Genetic Algorithm

Table 5.50 Selected parameters of the Cuckoo Search

Number of nests	Discovery rate of alien eggs	Levy exponent
50	0.25	1.00

Table 5.51 Selected parameters of the Particle Swarm Optimisation

Swarm size	Inertia weight	Inertia weight damping ratio	Personal learning coefficient	Global learning coefficient	Velocity limit coefficient
100	1.5	0.9	1.5	3.0	0.1

Table 5.52 Selected parameters of the traditional Genetic Algorithm

Population size	Crossover percentage	Mutation percentage	Extra range factor for crossover	Mutation rate	Selection method
50	0.5	0.5	0.4	0.1	Tournament

Table 5.53 Optimisation algorithm performance comparison

Run	Computing time (min)	Cuckoo Search			Particle Swarm Optimisation			Traditional Genetic Algorithm			Proposed Genetic Algorithm		
		Obj1	Obj2*	Obj3	Obj1	Obj2*	Obj3	Obj1	Obj2*	Obj3	Obj1	Obj2*	Obj3
1	180	6863.1	10.7 (0.9)	0.998759	7945.4	12.6 (1.8)	0.998773	6021.7	10.6 (1.4)	0.998718	5626.8	11.3 (0.9)	0.998732
2	180	6893.7	13.2 (1.9)	0.998776	8476.7	13.8 (1.5)	0.998763	6199.4	12.4 (1.4)	0.998708	5605.2	10.9 (0.8)	0.998708
3	180	6763.9	11.6 (0.7)	0.998759	8751.1	12.3 (1.6)	0.998849	5879.6	10.0 (0.5)	0.998870	5603.3	9.6 (0.5)	0.998804
4	180	6859.2	11.3 (0.7)	0.998772	8290.6	12.3 (1.0)	0.998792	5982.7	11.2 (0.6)	0.998741	5730.7	10.7 (0.7)	0.998798
5	180	6849.9	13.5 (2.0)	0.998769	8169.5	13.7 (1.6)	0.998764	5905.1	10.9 (1.2)	0.998732	5690.7	9.9 (0.8)	0.998787
Average:		6846.0	12.1 (1.2)	0.998767	8326.6	12.9 (1.5)	0.998788	5997.7	11.0 (1.1)	0.998754	5651.3	10.6 (0.8)	0.998766

**Note* This objective function has two parameters: mean value (standard deviation)

Table 5.54 The solution with the smallest cost (parts 1–2)

Product order	Electric bike component													Assembly agent
	C1	C2	C3	C4	C5	C6	C7	C8	C9	C10	C11	C12	C13	A
O1	2	2	2	2	2	5	2	2	5	5	2	-	2	1
O2	10	1	5	1	5	10	1	10	4	5	5	6	4	3
O3	7	4	4	1	10	9	3	7	10	1	10	3	6	3
O4	6	3	6	4	1	3	7	4	9	1	-	-	3	3
O5	1	1	4	3	6	10	6	6	6	4	6	7	10	3
O6	1	3	9	9	1	1	5	9	5	9	3	7	5	1

Table 5.55 The solution with the smallest cost (part 3)

	S1	S2	S3	S4	S5	S6	S7	S8	S9	S10	A
S1	0	0	0	0	0	0	0	0	0	0	0
S2	0	0	0	0	0	0	0	0	0	0	1
S3	0	0	0	0	0	0	0	0	0	0	0
S4	0	0	0	0	0	0	0	0	0	0	0
S5	0	0	0	0	0	0	0	0	0	0	1
S6	0	0	0	0	0	0	0	0	0	0	0
S7	0	0	0	0	0	0	0	0	0	0	0
S8	0	0	0	0	0	0	0	0	0	0	0
S9	0	0	0	0	0	0	0	0	0	0	0
S10	0	0	0	0	0	0	0	0	0	0	0

O1

	S1	S2	S3	S4	S5	S6	S7	S8	S9	S10	A
S1	0	0	0	0	0	0	0	0	0	0	1
S2	0	0	0	0	0	0	0	0	0	0	0
S3	0	0	0	0	0	0	0	0	0	0	0
S4	0	0	0	0	0	0	0	0	0	0	1
S5	1	0	0	0	0	0	0	0	0	0	0
S6	0	0	0	1	0	0	0	0	0	0	0
S7	0	0	0	0	0	0	0	0	0	0	0
S8	0	0	0	0	0	0	0	0	0	0	0
S9	0	0	0	0	0	0	0	0	0	0	0
S10	0	0	0	1	0	0	0	0	0	0	0

O2

	S1	S2	S3	S4	S5	S6	S7	S8	S9	S10	A
S1	0	0	0	0	0	0	0	0	0	0	1
S2	0	0	0	0	0	0	0	0	0	0	0
S3	1	0	0	0	0	0	0	0	0	0	0
S4	0	0	0	0	0	0	0	0	0	0	1
S5	0	0	0	0	0	0	0	0	0	0	0
S6	0	0	0	1	0	0	0	0	0	0	0
S7	1	0	0	0	0	0	0	0	0	0	0
S8	0	0	0	0	0	0	0	0	0	0	0
S9	1	0	0	0	0	0	0	0	0	0	0
S10	0	0	0	1	0	0	0	0	0	0	0

O3

	S1	S2	S3	S4	S5	S6	S7	S8	S9	S10	A
S1	0	0	0	0	0	0	0	0	0	0	1
S2	0	0	0	0	0	0	0	0	0	0	0
S3	1	0	0	0	0	0	0	0	0	0	0
S4	0	0	0	0	0	0	0	0	0	0	1
S5	0	0	0	0	0	0	0	0	0	0	0
S6	0	0	0	1	0	0	0	0	0	0	0
S7	1	0	0	0	0	0	0	0	0	0	0
S8	0	0	0	0	0	0	0	0	0	0	0
S9	1	0	0	0	0	0	0	0	0	0	0
S10	0	0	0	0	0	0	0	0	0	0	0

O4

	S1	S2	S3	S4	S5	S6	S7	S8	S9	S10	A
S1	0	0	0	0	0	0	0	0	0	0	1
S2	0	0	0	0	0	0	0	0	0	0	0
S3	1	0	0	0	0	0	0	0	0	0	0
S4	0	0	0	0	0	0	0	0	0	0	1
S5	0	0	0	0	0	0	0	0	0	0	0
S6	0	0	0	1	0	0	0	0	0	0	0
S7	1	0	0	0	0	0	0	0	0	0	0
S8	0	0	0	0	0	0	0	0	0	0	0
S9	0	0	0	0	0	0	0	0	0	0	0
S10	0	0	0	1	0	0	0	0	0	0	0

O5

	S1	S2	S3	S4	S5	S6	S7	S8	S9	S10	A
S1	0	0	0	0	0	0	0	0	0	0	1
S2	0	0	0	0	0	0	0	0	0	0	0
S3	1	0	0	0	0	0	0	0	0	0	0
S4	0	0	0	0	0	0	0	0	0	0	0
S5	0	0	0	0	0	0	0	0	0	0	1
S6	0	0	0	0	0	0	0	0	0	0	0
S7	1	0	0	0	0	0	0	0	0	0	0
S8	0	0	0	0	0	0	0	0	0	0	0
S9	1	0	0	0	0	0	0	0	0	0	0
S10	0	0	0	0	0	0	0	0	0	0	0

O6

Table 5.56 The solution with the smallest cost (part 4)

Assembly agent	Tier 1	Tier 2		Tier 3		Tier 4	
1	1	0	6	0	0	0	0
3	2	0	5	0	4	0	3

Table 5.57 Effect of the completion time threshold values on the cost of the products

Run	Computing time (min)	Maximum average completion time of the products		
		10 days	12 days	14 days
1	180	–	6616.6	5626.8
2	180	–	6006.8	5605.2
3	180	–	7000.4	5603.3
4	180	–	6190.1	5730.7
5	180	–	6277.3	5690.7
Average ($)		–	6418.2	5651.3

Table 5.58 Effect of the reliability threshold values on the cost of the products

Run	Computing time (min)	Defects per million opportunities (DPMO)		
		6200 (~4.0 sigma)	1300 (~4.5 sigma)	230 (~5.0 sigma)
1	180	5591.8	5626.8	–
2	180	5591.7	5605.2	–
3	180	5575.8	5603.3	–
4	180	5415.9	5730.7	–
5	180	5615.1	5690.7	–
Average ($)		5558.1	5651.3	–

5.5 Discussions

Since component release times, assembly times and shipping times in the proposed VCIM production scheduling model are stochastic parameters following the normal distribution, the completion times of the products are also stochastic. One question may be raised herein: what kind of distribution does the completion time of the products follow? To answer this question, a number of sample distributions of the completion times are visualised in Fig. 5.5. Generally speaking, the completion times of the products also follow the normal distribution which would be more clearly seen if the sample size of Monte Carlo simulation is increased. Running Monte Carlo simulation with a larger sample size will require more computing time.

There is no surprise that solving this case study problem took quite a long computing time, 180 min, as indicated in Tables 5.39 and 5.53. That is because Monte Carlo simulation is involved and this case study is a large-scale problem. It should be noted that the computer used herein is a very slow desktop computer, with processor: Intel(R) Core(TM)2 Quad CPU Q9400 @2.66 GHz and 4.00 GB RAM. From a decision maker's point of view, 180 min computing time is quite long. To overcome this, parallel computing techniques and super-computers can be used.

At the early stage of this research, the author attempted to solve several simplified versions of the case study problem, using three commercial optimisation solvers in Matlab, i.e. Pattern Search solver, Simulated Annealing solver and Genetic Algorithm solver. These solvers could work pretty well for the simplified problems. However, the solvers failed to solve this comprehensive problem. That is because the solvers required a massive computing time and their constraint

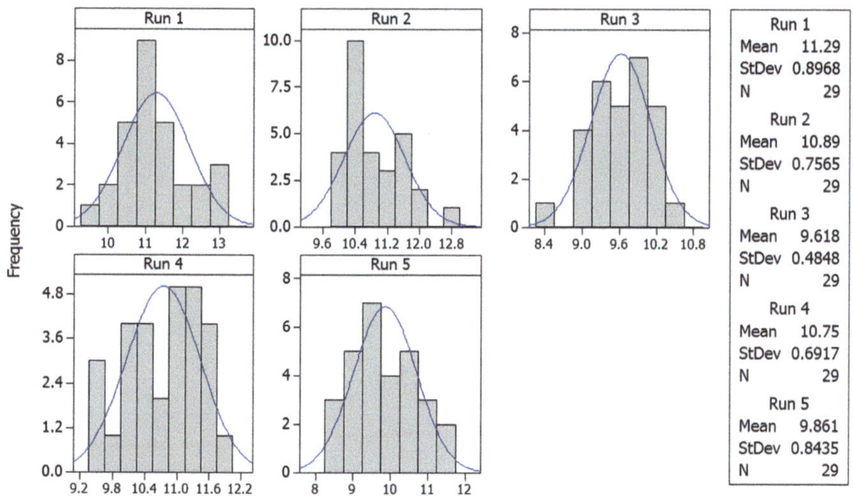

Fig. 5.5 Typical distributions of the completion times of the products

handling strategies were not efficient. Customised constraint handling strategies should be done, especially for large-scale problems.

When shifting from the deterministic model to the stochastic one, some chromosomes may be removed because they become infeasible due to new constraints in the stochastic model. The removed chromosomes are then replaced by randomly generated feasible chromosomes. Two typical transition scenarios from the deterministic model to the stochastic one are visualised in Figs. 5.6 and 5.7. In transition scenario in Fig. 5.6, the chromosome with the minimum total cost passed from the deterministic model to the stochastic one; while in transition scenario in Fig. 5.7 that kind of chromosome could not pass and it was removed and replaced (the minimum total cost was suddenly changed at the end of the evolutionary process).

Effect of the adaptive stop-and-restart-with-memory mechanism in the structure of the proposed GA can be clearly seen in Figs. 5.6 and 5.7. After about 700 generations, when the search process was about to converge to a local optimal solution, the proposed GA restarted the search process in which many completely new chromosomes were added while the top chromosomes were still kept. Thereby, the proposed GA has better capability to jump out of the local optima to reach the global optimal solution. The evidence here is that after about 700 generations, the minimum total cost still continued to go down significantly as shown in Figs. 5.6 and 5.7, which means that the proposed GA had overcome many local optima where the traditional optimisation algorithms are more likely to get stuck.

The convergence as shown in Figs. 5.6 and 5.7 also indicates that the proposed GA is capable of finding, at least, local optimal solutions to this large-scale case

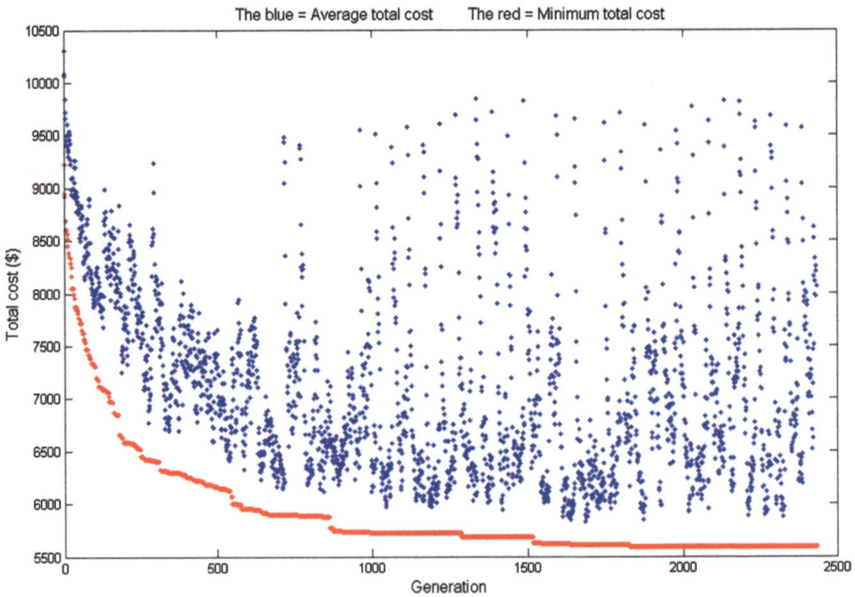

Fig. 5.6 Typical evolutionary process of the proposed GA (scenario 1)

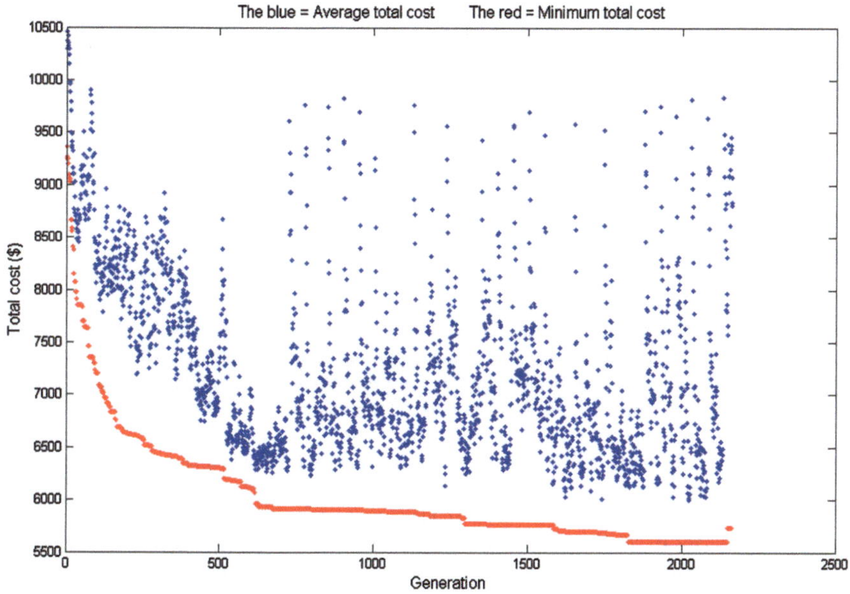

Fig. 5.7 Typical evolutionary process of the proposed GA (scenario 2)

study problem. It has a great potential to find the global optimal solution, because of the adaptive stop-and-restart-with-memory mechanism embedded in its structure. Nevertheless, proving and guaranteeing the global optimal solution to this problem is impossible because this is a large-scale NP-hard optimisation problem. From the decision makers' point of view, obtaining good or local optimal solutions with a reasonable computing time is acceptable.

5.6 Concluding Remarks

In this chapter, a comprehensive case study involving a VCIM system producing six types of electric bikes was carried out to demonstrate (1) the effectiveness of the proposed VCIM production scheduling model and (2) the efficiency of the proposed GA. The performance of the proposed VCIM production scheduling model was compared to those of two traditional scheduling models. The experimental data has shown that on average the proposed model could provide 6.4 and 5.3% better solutions, compared to two traditional models. In addition, the performance of the proposed GA has been compared to those of three popular optimisation algorithms in the literature. On average, solutions obtained by the proposed GA have 17.4, 32.1 and 5.8% better qualities, compared to solutions obtained by the Cuckoo Search, the Particle Swarm Optimisation and the traditional Genetic Algorithm, respectively. Finally, sensitivity analysis and some discussions were presented.

References

Australia_Post (2015). *Size & weight guidelines*. Australia Post. (http://auspost.com.au/parcels-mail/size-and-weight-guidelines.html?ilink=mm-size-and-weight-guidelines-1). Accessed on 6 Oct 2015.

Castanheira, I., Matos, A. S., Coelho, I., Gueifão, S., Roe, M., Calhau, M. A., et al. (2011). Six Sigma scale as a quality criterion for aggregation of food property measures. *Journal of Food Composition and Analysis, 24*(8), 1153–1159.

Heris, S. M. K. (2015a). Implementation of particle swarm optimization in MATLAB. Yarpiz. (http://yarpiz.com/440/ytea101-particle-swarm-optimization-pso-in-matlab-video-tutorial). Accessed on 2 Aug 2016.

Heris, S. M. K. (2015b). Implementation of real-coded genetic algorithm in MATLAB. Yarpiz. (http://yarpiz.com/23/ypea101-genetic-algorithms). Accessed on 1 Aug 2016.

Lair, N. A. M. (2008). *An integrated model for optimising manufacturing and distribution network scheduling* (Ph.D. thesis, School of Advance Manufacturing and Mechanical Engineering, University of South Australia).

Qing, H. H., & Xue, Z. S. (2009). A model for value-added e-marketplace provisioning: Case study from Alibaba.com. In C. Godart et al. (Eds.), *Software services for e-Business and e-Society* (Vol. 305, pp. 65–72). Berlin Heidelberg: Springer.

Saghaei, A., Najafi, H., & Noorossana, R. (2012). Enhanced rolled throughput yield: A new six sigma-based performance measure. *International Journal of Production Economics, 140*(1), 368–373.

Weinert, J., Ma, C., & Cherry, C. (2007). The transition to electric bikes in China: History and key reasons for rapid growth. *Transportation, 34*(3), 301–318.

Yang, X. S., & Deb, S. (2010). Engineering optimisation by Cuckoo Search. *International Journal of Mathematical Modelling and Numerical Optimisation, 1*(4), 330–343.

Chapter 6
Conclusions and Recommendations for Future Work

6.1 Introduction

In previous chapters, the literature review, the research gaps, the proposed VCIM production scheduling model and the proposed GA were presented. In addition, a comprehensive case study, i.e. a VCIM system producing various types of electric bikes was carried out to validate the effectiveness of the proposed VCIM production scheduling model as well as the proposed GA. In this chapter, conclusions, research contributions and recommendations for future work are presented.

6.2 Conclusions of the Research

VCIM is a relatively new concept, developed to help small and medium-size manufacturing enterprises survive and thrive in the current increasingly competitive global market. VCIM is still, conceptually, under development. Essentially, VCIM is a global manufacturing system which inherits two characteristics, namely "integration" and "temporary cooperation", from two mature systems, Computer-Integrated Manufacturing and Virtual Enterprise, respectively. A lot of work needs to be done to build a VCIM system, e.g. information sharing infrastructure, security, copyright, membership, quality control, effective production scheduling model and software, standardised components and products, shipment, payment, etc. Although some theoretical work has been done, a real VCIM system does not exist yet. However, inspired by the big success of Alibaba model (Liu and Avery 2009; Waite 2014), the researchers have a vision that VCIM systems will be the promising solutions for small and medium-size manufacturing enterprises in near future.

Production scheduling plays an important role in VCIM systems since it affects the quality, cost and lead time of products. Generally speaking, a VCIM system

S. D. Dao, *Modelling and Intelligent Optimisation of Production Scheduling in VCIM Systems*, Springer Theses, https://doi.org/10.1007/978-3-319-72113-2_6

works as follows. After receiving product orders, the VCIM system decomposes the orders into a number of components which can be independently produced. Then it does the VCIM production scheduling by (1) selecting some component suppliers to produce the required components, (2) selecting some assembly agents to assemble the required products and (3) scheduling the shipments to transport the components and products to the required destinations, so that a temporary VCIM production system can be formed to fulfil the product orders. Obviously, the VCIM production scheduling is critical to success of the system.

The literature review showed that there were two main limitations in the VCIM production scheduling, i.e. lack of a comprehensive scheduling model and lack of a robust optimisation solution method. This research aimed to overcome these two limitations. The outcome of this research would serve as an important foundation towards building as well as operating VCIM systems more effectively.

To overcome the limitation of existing models, an innovative production scheduling model for VCIM systems was developed herein. This is the most comprehensive VCIM production scheduling model to date, which is not only capable of supporting collaborative shipment scheduling and handling multiple product orders simultaneously but also dealing with multiple objective functions under uncertainties. The developed VCIM production scheduling model is fully integrated because all sub-problems, i.e. component supplier selection, assembly agent selection and collaborative shipment scheduling, are simultaneously solved for not only one but also multiple product orders. Solving these sub-problems one by one makes it impossible to achieve the global optimisation. The fully integrated VCIM production scheduling model is the only one that has the capability to facilitate the global optimisation.

It was demonstrated that GA is the best candidate for solving the optimisation problem in the proposed VCIM production scheduling model. The problem in the proposed scheduling model is a dynamic multi-objective optimisation problem under uncertainties. The problem is dynamic because the number of partners involved, i.e. the selected component suppliers and assembly agents, may vary from solution to solution. From computational complexity theory point of view, the problem in the proposed scheduling model is a NP-hard problem. According to He et al. (2012), there is no exact method that can find the global optimal solutions to NP-hard problems in polynomial time, and fast approximate heuristics and meta-heuristics are the popular approaches to search for high-quality/practical solutions. GA is one of the most popular meta-heuristics, often used to solve large-scale complex optimisation problems (Marian et al. 2012; Shahlaei et al. 2012). GA is capable of handling large search spaces, continuous/discrete problems and complex constraints (Fahimnia et al. 2008). That is why GA was chosen to solve the problem.

To overcome the limitation of existing optimisation methods for VCIM production scheduling and to solve the optimisation problem in the developed VCIM production scheduling model, an innovative GA was developed herein. It should be noted that GA is only a general search philosophy. There is no general GA capable of working best for every problem and the problem-specific customisation in chromosome encoding and genetic operations is always required to maximise the GA performance (Dao et al. 2014). In this research, an innovative GA with unique chromosome representation, modified genetic operations as well as novel algorithm structure was developed. To encode the solution, a special chromosome representation with four different parts representing four different sub-problems, namely component supplier selection, assembly agent selection, collaborative component shipment scheduling and collaborative product shipment scheduling, was developed. To better explore the solution space, three different types of crossover operations and four different types of mutation operations were also developed for the proposed GA. To enhance the global search capability, a novel GA structure with an adaptive stop-and-restart-with-memory mechanism was developed. And to maximise the GA performance, a systematic approach based on Taguchi experimental design was used to tune its parameters.

An effective strategy to include Monte Carlo simulation in a genetic algorithm was proposed herein. Since the developed VCIM production scheduling model is a stochastic model, Monte Carlo simulation is used to deal with the uncertainties. It is known that Monte Carlo simulation is very time-consuming, especially for the large-scale problems. Traditionally, Monte Carlo simulation is included in a genetic algorithm, from the beginning of the search process, which means Monte Carlo simulation is involved in all generations of the algorithm. This inclusion strategy is not very effective for solving large-scale problems, because it requires a massive computing time. To overcome this obstacle, a new inclusion strategy was proposed herein which includes Monte Carlo simulation in only a certain number of generations at the end of search process of the proposed GA. More specifically, in early generations the proposed GA worked with the deterministic version of the proposed VCIM production scheduling model where Monte Carlo simulation was not involved; in the late generations the proposed GA worked with the stochastic version of the model along with Monte Carlo simulation. The experimental data showed that the proposed strategy could provide much better solutions, on average 36.5% better, compared to the traditional strategy.

To demonstrate the effectiveness of the proposed VCIM production scheduling model and the efficiency of the proposed GA, a comprehensive case study involving a VCIM system producing six types of electric bikes was carried out where the performance of the proposed VCIM production scheduling model was compared to those of two traditional scheduling models. The experimental data showed that on average the proposed model provided 6.4 and 5.3% better solution, compared to two traditional models. In addition, performance of the proposed GA was compared to those of three popular optimisation algorithms in the literature. On average, solutions obtained by the proposed GA have 17.4, 32.1 and 5.8% better qualities, compared to solutions obtained by the Cuckoo Search, the Particle Swarm Optimisation and the traditional Genetic Algorithm, respectively.

6.3 Contributions of the Research

This research contributes to the body of knowledge in two major areas, namely VCIM systems and optimisation:

First, this research moves VCIM systems forward by developing a new VCIM production scheduling model. The developed model is the most comprehensive VCIM production scheduling model to date, which is not only capable of supporting collaborative shipment scheduling and handling multiple product orders simultaneously but also dealing with multiple objective functions, under uncertainties. As fully integrated, the developed model has a great capability to facilitate the global optimisation. In addition, the developed model can be applicable in other areas such as Job Shop Scheduling, Open Vehicle Routing, Project Planning and Scheduling, etc.

Second, this research contributes to the area of optimisation by developing a new GA, capable of solving complex large-scale optimisation problems. The developed GA has unique chromosome representation, modified genetic operations, novel algorithm structure and optimal parameter set. To encode the complex solution, a special chromosome representation with four different parts representing four different sub-problems is developed. To better explore the solution space, three different types of crossover operations and four different types of mutation operations are also developed. To enhance the global search capability, a novel GA structure with an adaptive stop-and-restart-with-memory mechanism is also developed here. To maximise the GA performance, a systematic approach based on Taguchi experimental design is used to tune its parameters. In addition, an effective strategy to include Monte Carlo simulation in a genetic algorithm is developed herein to reduce the computing time. The developed GA can be also applicable to solve many complex large-scale optimisation problems in various fields such as Supply Chain Management, Computer Science, Mathematics, Engineering, Physics, Biochemistry, Genetic/Molecular Biology, etc.

6.4 Recommendations for Future Work

Further work should address the following issues:

- Conducting more case studies to further demonstrate the effectiveness of the developed VCIM production scheduling model as well as the developed GA.
- Applying parallel computing techniques to shorten the computing time.
- Developing a more robust method to deal with multiple objective functions in the developed model.
- Applying Fuzzy Logic to deal with uncertainties in the model.
- Further generalising the developed VCIM production scheduling model by adding more realistic constraints and assumptions.

- Converting the developed VCIM production scheduling model and GA into a user-friendly decision support system that can help decision makers operate VCIM systems more easily and effectively.
- Further investigating in the technology transfer to real industrial environment to support implementing VCIM systems in real industry.
- Building a real VCIM system to test the proposed approach.

References

Dao, S. D., Abhary, K., & Marian, R. (2014). Optimisation of partner selection and collaborative transportation scheduling in virtual enterprises using GA. *Expert Systems with Applications, 41* (15), 6701–6717.

Fahimnia, B., Luong, L., & Marian, R. (2008). Optimization/simulation modeling of the integrated production-distribution plan: an innovative survey. *WSEAS Transactions on Business and Economics, 3*(5), 52–65.

He, K., Huang, W., & Jin, Y. (2012). An efficient deterministic heuristic for two-dimensional rectangular packing. *Computers & Operations Research, 39*(7), 1355–1363.

Liu, S., & Avery, M. (2009). *Alibaba: the inside story behind Jack Ma and the creation of the world's biggest online marketplace*. New York: HarperCollins e-books.

Marian, R.M., Luong, L. & Dao, S.D. (2012). Hybrid genetic algorithm optimisation of distribution networks—A comparative study, in SI Ao, O Castillo & X Huang (eds), *Intelligent Control and Innovative Computing*, Springer US, Boston, MA, pp. 109–122.

Shahlaei, M., Sobhani, A. M., Saghaie, L., & Fassihi, A. (2012). Application of an expert system based on Genetic Algorithm-Adaptive Neuro-Fuzzy Inference System (GA–ANFIS) in QSAR of cathepsin K inhibitors. *Expert Systems with Applications, 39*(6), 6182–6191.

Waite, S. (2014). Record-breaking Alibaba allocates to happy few. *FinanceAsia (Deal of the month: October)* 8–9. http://ezlibproxy.unisa.edu.au/login?, http://search.ebscohost.com/login.aspx?direct=true&db=bth&AN=99088865&site=ehost-live.

About the Author

Dr. Son Duy Dao was born in Thai Nguyen province, Vietnam. He obtained Bachelor's degree from Thai Nguyen University of Technology (TNUT), Vietnam in 2004. He received Master's degree and Ph.D. from University of South Australia, Australia in 2010 and 2017 respectively. He has worked as a lecturer for TNUT in Vietnam for a number of years. Currently, Dr. Dao is working as a postdoctoral research fellow at National University of Singapore, Singapore. His research interests are operations research, planning and scheduling optimisation, genetic algorithms, virtual computer-integrated manufacturing systems, supply chain management, and logistics. To date, he won a number of prestigious scholarships and research awards such as 322 Scholarship for master study in Australia, offered by Vietnamese Government (total value: $50,640), Australian Development Scholarship for Ph. D. study in Australia, offered by Australian Government (total value: $250,000), Australia Endeavour Award for Ph.D. study in Australia, offered by Australian Government (total value: $272,500), South Australia's International Student Award "*Academic Excellence in Postgraduate Research*", conferred by the Governor of South Australia, and Springer Prize for outstanding PhD thesis. Dr. Dao has published a number of journal articles, books, book chapters as well as conference papers. For his publications, it is advised to visit his Google Scholar Citations at: https://scholar.google.com.au/citations?user=WyM2f88AAAAJ&hl=en.

S. D. Dao, *Modelling and Intelligent Optimisation of Production Scheduling in VCIM Systems*, Springer Theses, https://doi.org/10.1007/978-3-319-72113-2

MIX
Papier aus verantwortungsvollen Quellen
Paper from responsible sources
FSC® C105338

If you have any concerns about our products,
you can contact us on
ProductSafety@springernature.com

In case Publisher is established outside the EU,
the EU authorized representative is:
Springer Nature Customer Service Center GmbH
Europaplatz 3, 69115 Heidelberg, Germany

Printed by Libri Plureos GmbH
in Hamburg, Germany